PRACTICE TESTS

for

LIVING
ENVIRONMENT
REGENTS

Answers Written By:

Charmian Foster

Science Chairperson – Johnson City High School

William Docekal

Science Teacher – Retired

Published by
TOPICAL REVIEW BOOK COMPANY
P. O. Box 328
Onsted, MI 49265-0328
E-mail: topicalrbc@aol.com • Website: www.topicalrbc.com

STUDENTS

To be successful on the Living Environment Regents you must be able to apply the concepts you have learned over the year. The exams and answers presented here provide you with about 300 questions that will test your understanding and your ability to apply your knowledge of biology. It is not enough to just do the practice exams before the Regents, you must be committed to seriously reviewing each answer and explanation until you feel confident of the concept.

Planning for the Regents begins perhaps a month or two months before the exam date. You do not want to wait until the last minute and cram. You should work a set of questions daily (about 15 to 20), going over the answers and reviewing the concepts involved. Star the questions you do not feel totally confident in and go back to those for more review and make notes in your margins!

If you work hard and do the exams carefully, review the answers and revisit areas of concern in a timely fashion, you should have success on the Regents.

LIVING ENVIRONMENT REGENTS

Published by
TOPICAL REVIEW BOOK COMPANY
P. O. Box 328
Onsted, MI 49265-0328

June 2007

Part A

Answer all questions in this part. [30]

Directions (1–30): For *each* statement or question, write in the space provided the *number* of the word or expression that, of those given, best completes the statement or answers the question.

1. Which statement describes a role of fungi in an ecosystem?
(1) They transfer energy to decaying matter.
(2) They release oxygen into the ecosystem.
(3) They recycle chemicals from dead organisms.
(4) They synthesize organic nutrients from inorganic substances.   1 _____

2. Which diagram best represents the levels of organization in the human body?

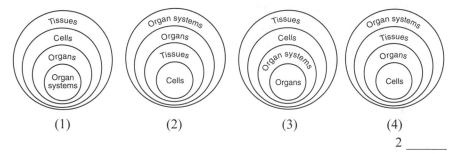

(1)   (2)   (3)   (4)
                                                                    2 _____

3. Which situation indicates that a disruption of homeostasis has taken place?
(1) the presence of hormones that keep the blood sugar level steady
(2) the maintenance of a constant body temperature
(3) cell division that is involved in normal growth
(4) a rapid rise in the number of red blood cells   3 _____

4. A protein on the surface of HIV can attach to proteins on the surface of healthy human cells. These attachment sites on the surface of the cells are known as
(1) receptor molecules   (3) molecular bases
(2) genetic codes   (4) inorganic catalysts   4 _____

5. Contractile vacuoles maintain water balance by pumping excess water out of some single-celled pond organisms. In humans, the kidney is chiefly involved in maintaining water balance. These facts best illustrate that
(1) tissues, organs, and organ systems work together to maintain homeostasis in all living things
(2) interference with nerve signals disrupts cellular communication and homeostasis within organisms
(3) a disruption in a body system may disrupt the homeostasis of a single-celled organism
(4) structures found in single-celled organisms can act in a manner similar to tissues and organs in multicellular organisms   5 _____

6. Which statement best explains the observation that clones produced from the same organism may *not* be identical?
(1) Events in meiosis result in variation.
(2) Gene expression can be influenced by the environment.
(3) Differentiated cells have different genes.
(4) Half the genetic information in offspring comes from each parent. 6 _____

7. A change in the base subunit sequence during DNA replication can result in
(1) variation within an organism
(2) rapid evolution of an organism
(3) synthesis of antigens to protect the cell
(4) recombination of genes within the cell   7 _____

8. The accompanying diagram represents a yeast cell that is in the process of budding, a form of asexual reproduction. Which statement describes the outcome of this process?

Nucleus
Bud
Nucleus

(1) The bud will develop into a zygote.
(2) The two cells that result will each contain half the species number of chromosomes.
(3) The two cells that result will have identical DNA.
(4) The bud will start to divide by the process of meiotic cell division. 8 _____

9. Two proteins in the same cell perform different functions. This is because the two proteins are composed of
(1) chains folded the same way and the same sequence of simple sugars
(2) chains folded the same way and the same sequence of amino acids
(3) chains folded differently and a different sequence of simple sugars
(4) chains folded differently and a different sequence of amino acids   9 _____

10. Even though each body cell in an individual contains the same DNA, the functions of muscle cells and liver cells are *not* the same because
(1) mutations usually occur in genes when muscle cells divide
(2) liver tissue develops before muscle tissue
(3) liver cells produce more oxygen than muscle cells
(4) liver cells use different genes than muscle cells   10 _____

11. The flounder is a species of fish that can live in very cold water. The fish produces an "antifreeze" protein that prevents ice crystals from forming in its blood. The DNA for this protein has been identified. An enzyme is used to cut and remove this section of flounder DNA that is then spliced into the DNA of a strawberry plant. As a result, the plant can now produce a protein that makes it more resistant to the damaging effects of frost. This process is known as
(1) sorting of genes   (3) recombination of chromosomes
(2) genetic engineering   (4) mutation by deletion of genetic material   11 _____

12. Some human body structures are represented in the accompanying diagram. In which structures would the occurrence of mutations have the greatest effect on human evolution?

 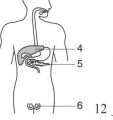

(1) 1 and 3   (3) 3 and 6
(2) 2 and 5   (4) 4 and 6   12 _____

13. A single pair of goldfish in an aquarium produced a large number of offspring. These offspring showed variations in body shape and coloration. The most likely explanation for these variations is that the
(1) offspring were adapting to different environments
(2) offspring were produced from different combinations of genes
(3) parent fish had not been exposed to mutagenic agents
(4) parent fish had not reproduced sexually   13 _____

14. A certain species has little genetic variation. The rapid extinction of this species would most likely result from the effect of
(1) successful cloning (3) environmental change
(2) gene manipulation (4) genetic recombination 14 _____

15. Which two structures of a frog would most likely have the same chromosome number?
(1) skin cell and fertilized egg cell (3) kidney cell and egg cell
(2) zygote and sperm cell (4) liver cell and sperm cell 15 _____

16. Tissues develop from a zygote as a direct result of the processes of
(1) fertilization and meiosis (3) mitosis and meiosis
(2) fertilization and differentiation (4) mitosis and differentiation 16 _____

17. The human female reproductive system is adapted for
(1) production of zygotes in ovaries
(2) external fertilization of gametes
(3) production of milk for a developing embryo
(4) transport of oxygen through a placenta to a fetus 17 _____

18. The letters in the accompanying diagram represent structures in a human female. Estrogen and progesterone increase the chance for successful fetal development by regulating activities within structure
(1) A (3) C
(2) B (4) D 18 _____

19. Which part of a molecule provides energy for life processes?
(1) carbon atoms (3) chemical bonds
(2) oxygen atoms (4) inorganic nitrogen 19 _____

20. Energy from organic molecules can be stored in ATP molecules as a direct result of the process of
(1) cellular respiration (3) diffusion
(2) cellular reproduction (4) digestion 20 _____

21. Which statement best describes how a vaccination can help protect the body against disease?
(1) Vaccines directly kill the pathogen that causes the disease.
(2) Vaccines act as a medicine that cures the disease.
(3) Vaccines cause the production of specific molecules that will react with and destroy certain microbes.
(4) Vaccines contain white blood cells that engulf harmful germs and prevent them from spreading throughout the body.   21 _____

22. The diagram below represents four different species of wild birds. Each species has feet with different structural adaptations.

| Mallard duck | Redheaded woodpecker | Northern cardinal | Common snipe |

The development of these adaptations can best be explained by the concept of
(1) inheritance of resistance to diseases that affect all these species
(2) inheritance of characteristics acquired after the birds hatched from the egg
(3) natural selection
(4) selective breeding   22 _____

23. The accompanying diagram represents a nucleus containing the normal chromosome number for a species. Which diagram bests illustrates the normal information of a cell that contains all of the genetic information needed for growth, development, and future reproduction of this species?

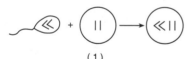
(1)

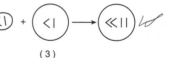
(3)

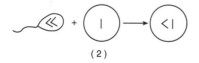
(2)

(4)   23 _____

24. The accompanying diagram represents events associated with a biochemical process that occurs in some organisms. Which statement concerning this process is correct?

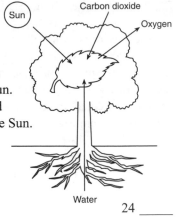

(1) The process represented is respiration and the primary source of energy for the process is the Sun.

√ (2) The process represented is photosynthesis and the primary source of energy for the process is the Sun.

(3) This process converts energy in organic compounds into solar energy which is released into the atmosphere.

(4) This process uses solar energy to convert oxygen into carbon dioxide.

24 _____

25. In the transfer of energy from the Sun to ecosystems, which molecule is one of the first to store this energy?

(1) protein   (2) fat   (3) DNA   √ (4) glucose

25 _____

26. The accompanying diagram represents two molecules that can interact with each other to cause a biochemical process to occur in a cell. Molecules *A* and *B* most likely represent

Molecule A   Molecule B

(1) a protein and a chromosome (3) a carbohydrate and an amino acid

√ (2) a receptor and a hormone (4) an antibody and a hormone

26 _____

27. The accompanying graph represents the amount of available energy at successive nutrition levels in a particular food web. The *X*s in the diagram represent the amount of energy that was most likely

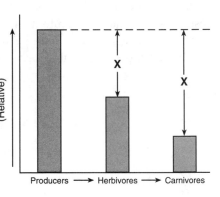

(1) changed into inorganic compounds

(2) retained indefinitely by the herbivores

(3) recycled back to the producers

√ (4) lost as heat to the environment

27 _____

28. The accompanying diagram represents an energy pyramid constructed from data collected from an aquatic ecosystem. Which statement best describes this ecosystem?

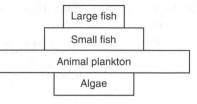

✓(1) The ecosystem is most likely unstable.
(2) Long-term stability of this ecosystem will continue.
(3) The herbivore populations will continue to increase in size for many years.
(4) The producer organisms outnumber the consumer organisms.   28 _____

29. In order to reduce consumption of nonrenewable resources, humans could
(1) burn coal to heat houses instead of using oil
✓(2) heat household water with solar radiation
(3) increase industrialization
(4) use a natural-gas grill to barbecue instead of using charcoal   29 _____

30. In 1859, a small colony of 24 rabbits was brought to Australia. By 1928 it was estimated that there were 500 million rabbits in a 1-million square mile section of Australia. Which statement describes a condition that probably contributed to the increase in the rabbit population?
(1) The rabbits were affected by many limiting factors.
(2) The rabbits reproduced by asexual reproduction.
(3) The rabbits were unable to adapt to the environment.
✓(4) The rabbits had no natural predators in Australia.   30 _____

Part B–1
Answer all questions in this part. [12]

Directions (31–42): For *each* statement or question, write in the space provided the *number* of the word or expression that, of those given, best completes the statement or answers the question.

31. What is the approximate length of the earthworm shown in the diagram below?

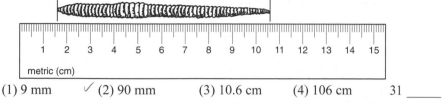

(1) 9 mm   ✓(2) 90 mm   (3) 10.6 cm   (4) 106 cm   31 _____

32. Information concerning the diet of crocodiles of different sizes is contained in the table below.

Percentage of Crocodiles of Different Lengths and Their Food Sources

Food Source	Group A 0.3–0.5 Meter	Group B 2.5–3.9 Meters	Group C 4.5–5.0 Meters
mammals	0	18	65
reptiles	0	17	48
fish	0	62	38
birds	0	17	0
snails	0	25	0
shellfish	0	5	0
spiders	20	0	0
frogs	35	0	0
insects	100	2	0

Which statement is *not* a valid conclusion based on the data?
(1) Spraying insecticides would have the most direct impact on group *A*.
(2) Overharvesting of fish could have a negative impact on group *C*.
√ (3) The smaller the crocodile is, the larger the prey.
(4) Group *B* has no preference between reptiles and birds.   32 _____

33. The accompanying diagram represents an incomplete section of a DNA molecule. The boxes represent unidentified bases. When the boxes are filled in, the total number of bases represented by the letter *A* (both inside and outside the boxes) will be
(1) 1   (2) 2   (3) 3   √ (4) 4   33 _____

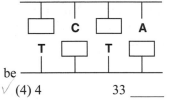

34. The accompanying graph shows the growth of a population of bacteria over a period of 80 hours. Which statement best describes section II of the graph?
(1) The population has reached the carrying capacity of the environment.
(2) The rate of reproduction is slower than in section I.
√ (3) The population is greater than the carrying capacity of the environment.
(4) The rate of reproduction exceeds the death rate.   34 _____

Growth of a Population of Bacteria

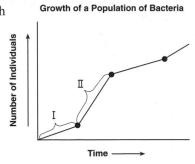

35. A classification system is shown in the table below.

Classification	Examples
Kingdom — animal	△, ○, ◁, ☆, ☐, ◇, ℰ, ▽
Phylum — chordata	△, ◁, ℰ, ☆, ☐
Genus — *Felis*	☐, ℰ
Species — *domestica*	☐

This classification scheme indicates that ☐ is most closely related to

☆   △   ☐   ℰ

(1)   (2)   (3)   (4) ✓   35 _____

36. Information concerning nests built in the same tree by two different bird species over a ten-year period is shown in the accompanying table.

Distance of Nest Above Ground (m)	Total Number of Nests Built by Two Different Species	
	A	B
less than 1	5	0
1–5	10	0
6–10	5	0
over 10	0	20

What inference best describes these two bird species?

✓ (1) They most likely do not compete for nesting sites because they occupy different niches.

(2) They do not compete for nesting sites because they have the same reproductive behavior.

(3) They compete for nesting sites because they build the same type of nest.

(4) They compete for nesting sites because they nest in the same tree at the same time.   36 _____

37. The diagram below shows the effect of spraying a pesticide on a population of insects over three generations.

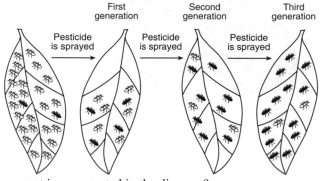

Which concept is represented in the diagram?

✓ (1) survival of the fittest   (3) succession

(2) dynamic equilibrium   (4) extinction   37 _____

38. In an ecosystem, the herring population was reduced by fishermen. As a result, the tuna, which feed on the herring, disappeared. The sand eels, which are eaten by herring, increased in number. The fishermen then overharvested the sand eel population. Cod and seabirds then decreased. Which food web best represents the feeding relationships in this ecosystem?

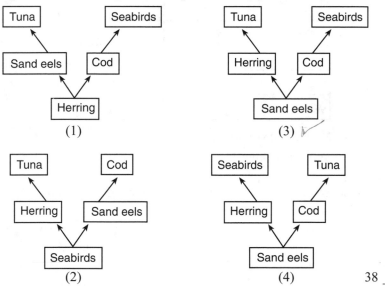

38 _____

Base your answers to questions 39 through 41 on the accompanying diagram, which represents systems in a human male and on your knowledge of biology.

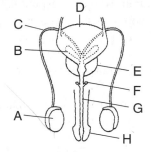

39. Which sequence represents the path of sperm leaving the body?

(1) $A \rightarrow C \rightarrow G$ (3) $E \rightarrow F \rightarrow H$

(2) $A \rightarrow C \rightarrow B$ (4) $D \rightarrow F \rightarrow G$ 39 _____

40. Which structures aid in the transport of sperm by secreting fluid?

(1) A and H (2) B and E (3) C and D (4) D and H 40 _____

41. Which structure has both reproductive and excretory functions?

(1) A (2) G (3) C (4) D 41 _____

42. Two food chains are represented below.

Food chain A: aquatic plant→insect→frog→hawk
Food chain B: grass → rabbit → hawk

Decomposers are important for supplying energy for

(1) food chain A, only   (3) both food chain A and food chain B

(2) food chain B, only   (4) neither food chain A nor food chain B 42 _____

Part B–2
Answer all questions in this part. [13]

Directions (43–55): For those questions that are followed by four choices, record in the space provided the *number* of the choice that, of those given, best completes the statement or answers the question. For all other questions in this part, follow the directions given in the question and record your answers in the spaces provided.

Base your answers to questions 43 through 45 on the accompanying diagrams and on your knowledge of biology. The diagrams represent two different cells and some of their parts. The diagrams are not drawn to scale.

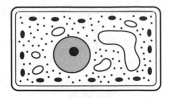

Cell A

43. Identify an organelle in cell *A* that is the site of autotrophic nutrition. [1]

_____

44. Identify the organelle labeled *X* in cell *B*. [1]

_____

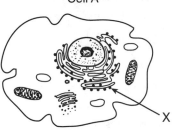

X

Cell B

45. Which statement best describes these cells?
(1) Cell *B* lacks vacuoles while cell *A* has them.
(2) DNA would not be found in either cell *A* or cell *B*.
(3) Both cell *A* and cell *B* use energy released from ATP.
(4) Both cell *A* and cell *B* produce antibiotics.

45 _____

Base your answers to questions 46 through 48 on the accompanying diagram and on your knowledge of biology.

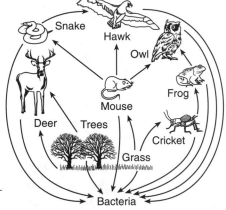

46. What is an appropriate title for this diagram?
(1) Energy Flow in a Community
(2) Ecological Succession
(3) Biological Evolution
(4) A Food Chain   46 _____

47. Which organism carries out autotrophic nutrition?
(1) hawk   (2) cricket   (3) grass   (4) deer   47 _____

48. State what would most likely happen to the cricket population if all of the grasses were removed. [1]

_____

_____

Base your answers to questions 49 through 53 on the information and diagrams below and on your knowledge of biology.

The laboratory setups represented below were used to investigate the effect of temperature on cellular respiration in yeast (a single-celled organism). Each of two flasks containing equal amounts of a yeast-glucose solution was submerged in a water bath, one kept at 20°C and one kept at 35°C. The number of gas bubbles released from the glass tube in each setup was observed and the results were recorded every 5 minutes for a period of 25 minutes. The data are summarized in the table below.

June 2007

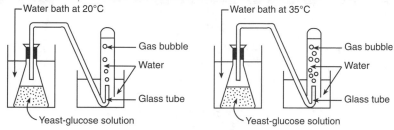

Data Table

Time	Total Number of Bubbles Released	
(minutes)	20°C	35°C
5	0	5
10	5	15
15	15	30
20	30	50
25	45	75

Directions (49–51): Using the information in the data table, construct a line graph on the grid on the next page, following the directions below.

49. Mark an appropriate scale on each axis. [1]

50. Plot the data for the total number of bubbles released at 20°C on the grid on the next page. Surround each point with a small circle and connect the points. [1]

Example: ⊙—⊙—⊙

51. Plot the data for the total number of bubbles released at 35°C on the grid. Surround each point with a small triangle and connect the points. [1]

Example: △—△—△

**The Effect of Temperature
on Respiration in Yeast**

Total Number of Bubbles Released

Time (minutes)

52. State *one* relationship between temperature and the rate of gas production in yeast. [1]

_____

_____

53. Identify the gas that would be produced by the process taking place in both laboratory setups. [1]_____

Base your answers to questions 54 and 55 on the accompanying diagram and on your knowledge of biology.

54. Identify the organ labeled *X*. [1]

_____

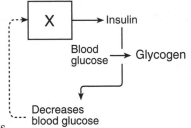

55. The dashed line in the diagram represents

(1) a digestive process   (3) cellular differentiation

(2) a feedback mechanism   (4) recycling of organic chemicals   55 _____

Part C
Answer all questions in this part. [17]
Directions (56–61): **Record your answers in the spaces provided.**

56. An experiment was carried out to determine how competition for living space affects plant height. Different numbers of plants were grown in three pots, *A*, *B*, and *C*. All three pots were the same size. The data collected are shown in the table below.

Average Daily Plant Height (mm)							
	Day 1	Day 2	Day 3	Day 4	Day 5	Day 6	Day 7
Pot A—5 plants	2	4	6	8	10	14	16
Pot B—10 plants	2	4	6	8	10	12	12
Pot C—20 plants	2	2	2	6	6	8	8

Analyze the experiment that produced the data shown in the table.
In your answer be sure to:
- state a hypothesis for the experiment [1]
- identify *one* factor, other than pot size, that should have been kept the same in each experimental group [1]
- identify the dependent variable [1]
- state whether the data supports or fails to support your hypothesis and justify your answer [1]

_____

_____

_____

_____

_____

_____

_____

_____

_____

_____

_____

_____

_____

_____

57. In many investigations, both in the laboratory and in natural environments, the pH of substances is measured. Explain why pH is important to living things. In your explanation be sure to:
- identify *one* example of a life process of an organism that could be affected by a pH change [1]
- state *one environmental* problem that is directly related to pH [1]
- identify *one* possible cause of this environmental problem [1]

_____

_____

_____

_____

_____

_____

_____

_____

_____

_____

_____

_____

_____

_____

Base your answer to question 58 on the information below and on your knowledge of biology.

Cargo ships traveling to the Great Lakes from the Caspian Sea in Eurasia often carry water in tanks known as ballast tanks. This water helps the ships to be more stable while crossing the ocean. Upon arrival in the Great Lakes, this water is pumped out of the ships. Often this water contains species that are not native to the Great Lakes environment. The zebra mussel is one species that was introduced into the Great Lakes in this way.

Although large numbers of zebra mussels often clog water intake pipes of power plants and other industries, the mussels have a benefit. Each mussel filters about a quart of water per day, absorbing cancer-causing PCB's from lake water in the process.

The goby, a bottom-feeding fish from Europe, was introduced into the Great Lakes in a similar way a few years later. The gobies have become a dominant species in the Great Lakes, eating small zebra mussels and the eggs and young of other fish. Gobies are eaten by large sport fish. These sport fish have been tested and PCB's have been found in their tissues. Recommendations have been made that people limit the number of sport fish they eat.

58. Explain how the introduction of foreign species can often cause environmental problems. In your answer be sure to:
- state how the zebra mussels and gobies were introduced into the United States [1]
- state *one* way either the zebra mussels *or* gobies have become a problem in their new environment [1]
- describe how *both* zebra mussels and gobies contribute to increasing the concentration of PCB's in sport fish [2]

_____

_____

_____

_____

_____

_____

_____

_____

_____

_____

_____

_____

_____

59. Knowledge of human genes gained from research on the structure and function of human genetic material has led to improvements in medicine and health care for humans.
- state *two* ways this knowledge has improved medicine and health care for humans [2]
- identify *one* specific concern that could result from the application of this knowledge [1]

_____

_____

_____

_____

_____

_____

Base your answers to questions 60 and 61 on the information below and on your knowledge of biology.

You are the owner of a chemical company. Many people in your community have been complaining that rabbits are getting into their gardens and eating the flowering plants and vegetables they have planted. Your company is developing a new chemical product called Bunny Hop-Away that repels rabbits. This product would be sprayed on the plants to prevent the rabbits from eating them. Certain concerns need to be considered before you make the product available for public use.

60. State *two* environmental concerns that should be considered before the product is sold and used by the public. [2]

_____

_____

_____

_____

_____

61. State *one* safety procedure that should be followed when the product is sprayed on plants. [1]

_____

_____

Part D
Answer all questions in this part. [13]
Directions (62–73): For those questions that are followed by four choices, circle the *number* of the choice, that, of those given, best completes the statement or answers the question. For all other questions in this part, follow the directions given in the questions and record your answers in the spaces provided.

62. Students were asked to determine if they could squeeze a clothespin more times in a minute after resting than after exercising. An experiment that accurately tests this question should include all of the following *except*
(1) a hypothesis on which to base the design of the experiment
(2) a large number of students
(3) two sets of clothespins, one that is easy to open and one that is more difficult to open
(4) a control group and an experimental group with equal numbers of students of approximately the same age   62 _____

Base your answers to questions 63 through 65 on the accompanying diagram and on your knowledge of biology. The diagram shows the results of a technique used to analyze DNA.

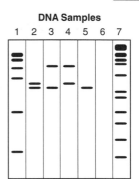

63. This technique used to analyze DNA directly results in
(1) synthesizing large fragments of DNA
(2) separating DNA fragments on the basis of size
(3) producing genetically engineered DNA molecules
(4) removing the larger DNA fragments from the samples   63 _____

64. This laboratory technique is known as
(1) gel electrophoresis   (3) protein synthesis
(2) DNA replication   (4) genetic recombination   64 _____

65. State *one* specific way the results of this laboratory technique could be used. [1]

_____

_____

66. Which statement best describes a controlled experiment?
(1) It eliminates the need for dependent variables.
(2) It shows the effect of a dependent varialbe on an independent variable.
(3) It avoids the use of variables.
(4) It tests the effect of a single independent variable.   66 _____

67. Which statement best describes a change that usually takes place in the human body when the heart rate increases as a result of exercise?
(1) More oxygen is delivered to muscle cells.
(2) Blood cells are excreted at a faster rate.
(3) The rate of digestion increases.
(4) No hormones are produced.   67 _____

68. The cactus finch, warbler finch, and woodpecker finch all live on one island. Based on the information in the accompanying diagram, which one of these finches is *least* likely to compete with the other two for food? Support your answer with an explanation. [1]

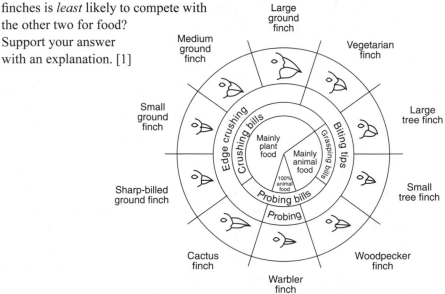

From: *Galapagos: A Natural History Guide*

Variations in Beaks of Galapagos Islands Finches

_____

_____

_____

Base your answers to questions 69 and 70 on the information below and on your knowledge of biology.

Evolutionary changes have been observed in beak size in a population of medium ground finches in the Galapagos Islands. Given a choice of small and large seeds, the medium ground finch eats mostly small seeds, which are easier to crush. However, during dry years, all seeds are in short supply. Small seeds are quickly consumed, so the birds are left with a diet of large seeds. Studies have shown that this change in diet may be related to an increase in the average size of the beak of the medium ground finch.

69. The most likely explanation for the increase in average beak size of the medium ground finch is that the
(1) trait is inherited and birds with larger beaks have greater reproductive success
(2) birds acquired larger beaks due to the added exercise of feeding on large seeds
(3) birds interbred with a larger-beaked species and passed on the trait
(4) lack of small seeds caused a mutation which resulted in a larger beak   69 _____

70. In exceptionally dry years, what most likely happens in a population of medium ground finches?
(1) There is increased cooperation between the birds.
(2) Birds with large beaks prey on birds with small beaks.
(3) The finches develop parasitic relationships with mammals.
(4) There is increased competition for a limited number of small seeds. 70 _____

71. Cell *A* shown below is a typical red onion cell in water on a slide viewed with a compound light microscope. Draw a diagram of how cell *A* would most likely look after salt water has been added to the slide and label the cell membrane in your di

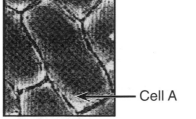

Cell A

Base your answers to questions 72 and 73 on the diagram below and on your knowledge of biology. The diagram shows the changes that occurred in a beaker after 30 minutes. The beaker contained water, food coloring, and a bag made from dialysis tubing membrane.

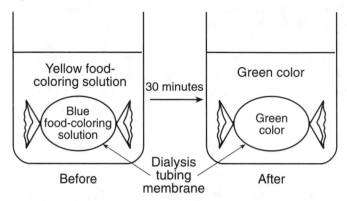

72. When the colors yellow and blue are combined, they produce a green color. Which statement most likely describes the relative sizes of the yellow and blue food-coloring molecules in the diagram?
(1) The yellow food-coloring molecules are small, while the blue food-coloring molecules are large.
(2) The yellow food-coloring molecules are large, while the blue food-coloring molecules are small.
(3) Both the yellow food-coloring molecules and the blue food-coloring molecules are large.
(4) Both the yellow food-coloring molecules and the blue food-coloring molecules are small.   72 _____

73. Which statement best explains the changes shown?
(1) Molecular movement was aided by the presence of specific carbohydrate molecules on the surface of the membrane.
(2) Molecular movement was aided by the presence of specific enzyme molecules on the surface of the membrane.
(3) Molecules moved across the membrane without additional energy being supplied.
(4) Molecules moved across the membrane only when additional energy was supplied.   73 _____

June 2008
Part A
Answer all questions in this part. [30]

Directions (1–30): For *each* statement or question, write in the space provided the *number* of the word or expression that, of those given, best completes the statement or answers the question.

1. The accompanying chart contains both autotrophic and heterotrophic organisms. Organisms that carry out only heterotrophic nutrition are found in

A	owl	cat	shark
B	mouse	corn	dog
C	squirrel	bluebird	alga

(1) row *A*, only   (3) rows *A* and *B*
(2) row *B*, only   (4) rows *A* and *C*   1 _____

2. A stable pond ecosystem would *not* contain
(1) materials being cycled   (3) decomposers
(2) oxygen   (4) more consumers than producers   2 _____

3. Although all of the cells of a human develop from one fertilized egg, the human is born with many different types of cells. Which statement best explains this observation?
(1) Developing cells may express different parts of their identical genetic instructions.
(2) Mutations occur during development as a result of environmental conditions.
(3) All cells have different genetic material.
(4) Some cells develop before other cells.   3 _____

4. Humans require organ systems to carry out life processes. Single-celled organisms do not have organ systems and yet they are able to carry out life processes. This is because
(1) human organ systems lack the organelles found in single-celled organisms
(2) a human cell is more efficient than the cell of a single-celled organism
(3) it is not necessary for single-celled organisms to maintain homeostasis
(4) organelles present in single-celled organisms act in a manner similar to organ systems   4 _____

5. Certain poisons are toxic to organisms because they interfere with the function of enzymes in mitochondria. This results directly in the inability of the cell to
(1) store information   (3) release energy from nutrients
(2) build proteins   (4) dispose of metabolic wastes   5 _____

6. At warm temperatures, a certain bread mold can often be seen growing on bread as a dark-colored mass. The same bread mold growing on bread in a cooler environment is red in color. Which statement most accurately describes why this change in the color of the bread mold occurs?

(1) Gene expression can be modified by interactions with the environment.
(2) Every organism has a different set of coded instructions.
(3) The DNA was altered in response to an environmental condition.
(4) There is no replication of genetic material in the cooler environment.   6 _____

7. Asexually reproducing organisms pass on hereditary information as

(1) sequences of A, T, C, and G   (3) folded protein molecules
(2) chains of complex amino acids   (4) simple inorganic sugars   7 _____

8. Species of bacteria can evolve more quickly than species of mammals because bacteria have

(1) less competition   (3) lower mutation rates
(2) more chromosomes   (4) higher rates of reproduction   8 _____

9. The diagram below represents the synthesis of a portion of a complex molecule in an organism.

Building blocks   Product

Which row in the chart could be used to identify the building blocks and product in the diagram?

Row	Building Blocks	Product
(1)	starch molecules	glucose
(2)	amino acid molecules	part of protein
(3)	sugar molecules	ATP
(4)	DNA molecules	part of starch

9 _____

10. Which diagram best represents the relative locations of the structures in the list below?   A–chromosome   B–nucleus   C–cell   D–gene

(1)   (2)   (3)   (4)   10 _____

11. Which nuclear process is represented below?

| A DNA molecule untwists. | → | The two strands of DNA separate. | → | Molecular bases pair up. | → | Two identical DNA molecules are produced. |

(1) recombination   (3) replication

(2) fertilization   (4) mutation   11 _____

12. For centuries, certain animals have been crossed to produce offspring that have desirable qualities. Dogs have been mated to produce Labradors, beagles, and poodles. All of these dogs look and behave very differently from one another. This technique of producing organisms with specific qualities is known as

(1) gene replication   (3) random mutation

(2) natural selection   (4) selective breeding   12 _____

13. Certain insects resemble the bark of the trees on which they live. Which statement provides a possible biological explanation for this resemblance?

(1) The insects needed camouflage so they developed protective coloration.

(2) Natural selection played a role in the development of this protective coloration.

(3) The lack of mutations resulted in the protective coloration.

(4) The trees caused mutations in the insects that resulted in protective coloration.   13 _____

14. When is extinction of a species most likely to occur?

(1) when environmental conditions remain the same and the proportion of individuals within the species that lack adaptive traits increases

(2) when environmental conditions remain the same and the proportion of individuals within the species that possess adaptive traits increases

(3) when environmental conditions change and the adaptive traits of the species favor the survival and reproduction of some of its members

(4) when environmental conditions change and the members of the species lack adaptive traits to survive and reproduce   14 _____

15. In what way are photosynthesis and cellular respiration similar?

(1) They both occur in chloroplasts.

(2) They both require sunlight.

(3) They both involve organic and inorganic molecules.

(4) They both require oxygen and produce carbon dioxide.   15 _____

16. Which process will increase variations that could be inherited?
(1) mitotic cell division (3) recombination of genes
(2) active transport (4) synthesis of proteins 16 _____

17. Some cells involved in the process of reproduction are represented in the accompanying diagram. The process of meiosis formed

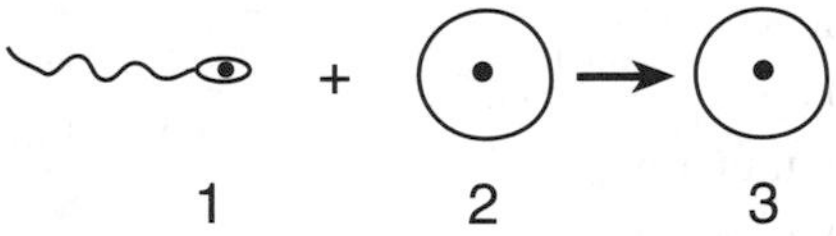

1   2   3

(1) cell 1, only (2) cells 1 and 2 (3) cell 3, only (4) cells 2 and 3 17 _____

18. Kangaroos are mammals that lack a placenta. Therefore, they must have an alternate way of supplying the developing embryo with
(1) nutrients (3) enzymes
(2) carbon dioxide (4) genetic information 18 _____

19. Which substance is the most direct source of the energy that an animal cell uses for the synthesis of materials?
(1) ATP (2) glucose (3) DNA (4) starch 19 _____

20. To increase chances for a successful organ transplant, the person receiving the organ should be given special medications. The purpose of these medications is to
(1) increase the immune response in the person receiving the transplant
(2) decrease the immune response in the person receiving the transplant
(3) decrease mutations in the person receiving the transplant
(4) increase mutations in the person receiving the transplant 20 _____

21. The diagram below represents the cloning of a carrot plant.

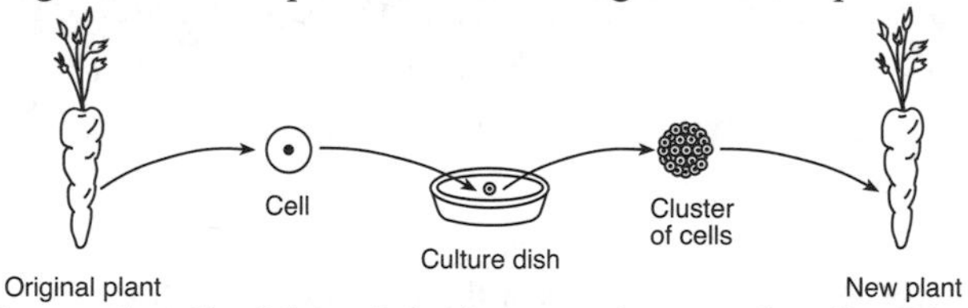

Original plant   Cell   Culture dish   Cluster of cells   New plant

Compared to each cell of the original carrot plant, each cell of the new plant will have
(1) the same number of chromosomes and the same types of genes
(2) the same number of chromosomes, but different types of genes
(3) half the number of chromosomes and the same types of genes
(4) half the number of chromosomes, but different types of genes 21 _____

22. The development of an embryo is represented in the accompanying diagram. These changes in the form of the embryo are a direct result of

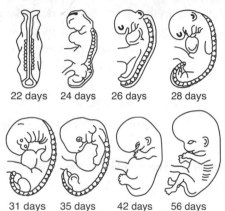

22 days   24 days   26 days   28 days

31 days   35 days   42 days   56 days

(Not drawn to scale)

(1) uncontrolled cell division and mutations
(2) differentiation and growth
(3) antibodies and antigens inherited from the father
(4) meiosis and fertilization

22 _____

23. The accompanying diagram represents an event that occurs in the blood. Which statement best describes this event?

Cell A

(1) Cell A is a white blood cell releasing antigens to destroy bacteria.
(2) Cell A is a cancer cell produced by the immune system and it is helping to prevent disease.
(3) Cell A is a white blood cell engulfing diseasecausing organisms.
(4) Cell A is protecting bacteria so they can reproduce without being destroyed by predators.

23 _____

24. In an ecosystem, the growth and survival of organisms are dependent on the availability of the energy from the Sun. This energy is available to organisms in the ecosystem because
(1) producers have the ability to store energy from light in organic molecules
(2) consumers have the ability to transfer chemical energy stored in bonds to plants
(3) all organisms in a food web have the ability to use light energy
(4) all organisms in a food web feed on autotrophs

24 _____

25. Which factor has the greatest influence on the type of ecosystem that will form in a particular geographic area?
(1) genetic variations in the animals
(2) climate conditions
(3) number of carnivores
(4) percentage of nitrogen gas in the atmosphere

25 _____

26. Farming reduces the natural biodiversity of an area, yet farms are necessary to feed the world's human population. This situation is an example of

(1) poor land use　　　　　(3) conservation
(2) a trade-off　　　　　　(4) a technological fix　　　　　26 ____

27. A food chain is represented below.

Grass → Cricket → Frog → Owl

This food chain contains

(1) 4 consumers and no producers
(2) 1 predator, 1 parasite, and 2 producers
(3) 2 carnivores and 2 herbivores
(4) 2 predators, 1 herbivore, and 1 producer　　　　　27 ____

28. A volcanic eruption destroyed a forest, covering the soil with volcanic ash. For many years, only small plants could grow. Slowly, soil formed in which shrubs and trees could grow. These changes are an example of

(1) manipulation of genes　(3) ecological succession
(2) evolution of a species　(4) equilibrium　　　　　28 ____

29. A major reason that humans can have such a significant impact on an ecological community is that humans

(1) can modify their environment through technology
(2) reproduce faster than most other species
(3) are able to increase the amount of finite resources available
(4) remove large amounts of carbon dioxide from the air　　　　　29 ____

30. Rabbits are herbivores that are not native to Australia. Their numbers have increased steadily since being introduced into Australia by European settlers. One likely reason the rabbit population was able to grow so large is that the rabbits

(1) were able to prey on native herbivores
(2) reproduced more slowly than the native animals
(3) successfully competed with native herbivores for food
(4) could interbreed with the native animals　　　　　30 ____

Part B–1

Answer all questions in this part. [12]

Directions (31–42): For *each* statement or question, write in the space provided the *number* of the word or expression that, of those given, best completes the statement or answers the question.

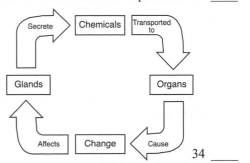

31. Which laboratory procedure is represented in the accompanying diagram?
(1) placing a coverslip over a specimen
(2) removing a coverslip from a slide
(3) adding stain to a slide without removing the coverslip
(4) reducing the size of air bubbles under a coverslip   31 ____

32. In the United States, there has been relatively little experimentation involving the insertion of genes from other species into human DNA. One reason for the lack of these experiments is that
(1) the subunits of human DNA are different from the DNA subunits of other species
(2) there are many ethical questions to be answered before inserting foreign genes into human DNA
(3) inserting foreign DNA into human DNA would require using techniques completely different from those used to insert foreign DNA into the DNA of other mammals
(4) human DNA always promotes human survival, so there is no need to alter it   32 ____

33. The development of an experimental research plan should *not* include a
(1) list of safety precautions for the experiment
(2) list of equipment needed for conducting the experiment
(3) procedure for the use of technologies needed for the experiment
(4) conclusion based on data expected to be collected in the experiment 33 ____

34. The accompanying diagram represents an interaction between parts of an organism. The term *chemicals* in this diagram represents
(1) starch molecules
(2) DNA molecules
(3) hormone molecules
(4) receptor molecules   34 ____

35. A student performed an experiment to demonstrate that a plant needs chlorophyll for photosynthesis. He used plants that had green leaves with white areas. After exposing the plants to sunlight, he removed a leaf from each plant and processed the leaves to remove the chlorophyll. He then tested each leaf for the presence of starch. Starch was found in the area of the leaf that was green, and no starch was found in the area of the leaf that was white. He concluded that chlorophyll is necessary for photosynthesis. Which statement represents an assumption the student had to make in order to draw this conclusion?
(1) Starch is synthesized from the glucose produced in the green areas of the leaf.
(2) Starch is converted to chlorophyll in the green areas of the leaf.
(3) The white areas of the leaf do not have cells.
(4) The green areas of the leaf are heterotrophic.   35 ____

36. The accompanying diagram represents two cells, X and Y. Which statement is correct concerning the structure labeled A?

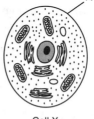 

Cell X   Cell Y

(1) It aids in the removal of metabolic wastes in both cell X and cell Y.
(2) It is involved in cell communication in cell X, but not in cell Y.
(3) It prevents the absorption of CO_2 in cell X and O_2 in cell Y.
(4) It represents the cell wall in cell X and the cell membrane in cell Y.   36 ____

37. The accompanying graph provides information about the reproductive rates of four species of bacteria, A, B, C, and D, at different temperatures. Which statement is a valid conclusion based on the information in the graph?

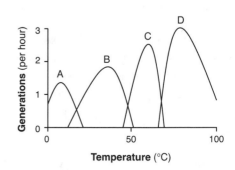

(1) Changes in temperature cause bacteria to adapt to form new species.
(2) Increasing temperatures speed up bacterial reproduction.
(3) Bacteria can survive only at temperatures between 0°C and 100°C.
(4) Individual species reproduce within a specific range of temperatures.   37 ____

38. The accompanying diagram shows some of the steps in protein synthesis. The section of DNA being used to make the strand of mRNA is known as a

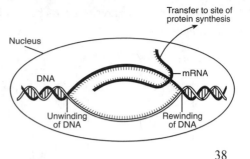

(1) carbohydrate
(2) gene
(3) ribosome
(4) chromosome

38 _____

39. An energy pyramid is shown to the right. Which graph best represents the relative energy content of the levels of this pyramid?

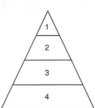

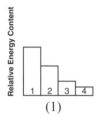
(1)

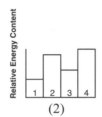
(2)

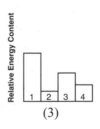
(3)

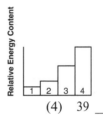
(4)   39 _____

40. The diagram below represents four different species of bacteria.

Species A	Species B	Species C	Species D

Which statement is correct concerning the chances of survival for these species if there is a change in the environment?
(1) Species A has the best chance of survival because it has the most genetic diversity.
(2) Species C has the best chance of survival because it has no gene mutations.
(3) Neither species B nor species D will survive because they compete for the same resources.
(4) None of the species will survive because bacteria reproduce asexually.

40 _____

41. The diagram below represents possible evolutionary relationships between groups of organisms.

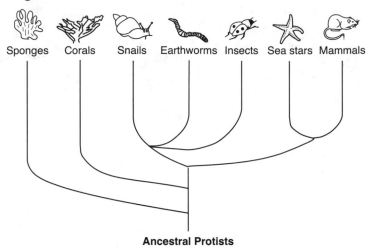

Ancestral Protists

Which statement is a valid conclusion that can be drawn from the diagram?
(1) Snails appeared on Earth before corals.
(2) Sponges were the last new species to appear on Earth.
(3) Earthworms and sea stars have a common ancestor.
(4) Insects are more complex than mammals.   41 ____

42. On which day did the population represented in the graph below reach the carrying capacity of the ecosystem?

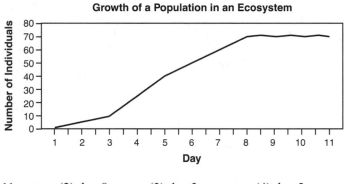

(1) day 11   (2) day 8   (3) day 3   (4) day 5   42 ____

Part B–2
Answer all questions in this part. [13]

Directions **(43–55): For those questions that are followed by four choices, write in the space provided the *number* of the choice that, of those given, best completes the statement or answers the question. For all other questions in this part, follow the directions given in the question and record your answers in the spaces provided.**

Base your answers to questions 43 through 47 on the information below and on your knowledge of biology.

Each year, a New York State power agency provides its customers with information about some of the fuel sources used in generating electricity. The accompanying table applies to the period of 2002–2003.

Fuel Sources Used

Fuel Source	Percentage of Electricity Generated
hydro (water)	86
coal	5
nuclear	4
oil	1
solar	0

Directions (43 and 44): Using the information given, construct a bar graph on the grid, following the directions below.

43. Mark an appropriate scale on the axis labeled "Percentage of Electricity Generated." [1]

44. Construct vertical bars to represent the data. Shade in *each* bar. [1]

45. Identify *one* fuel source in the table that is considered a fossil fuel. [1]

_____

46. Identify *one* fuel source in the table that is classified as a renewable resource. [1]

_____

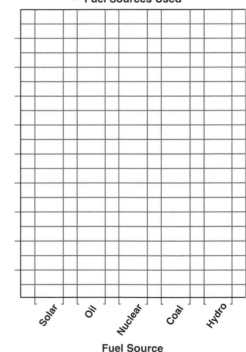

Fuel Sources Used

Percentage of Electricity Generated

Solar   Oil   Nuclear   Coal   Hydro

Fuel Source

47. State *one* specific environmental problem that can result from burning coal to generate electricity. [1]

_____

_____

Base your answers to questions 48 and 49 on the accompanying diagram that shows some interactions between several organisms located in a meadow environment and on your knowledge of biology.

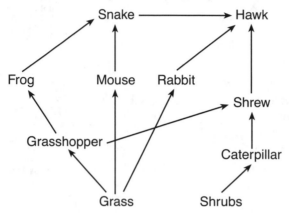

48. A rapid *decrease* in the frog population results in a change in the hawk population. State how the hawk population may change. Support your answer. [1]

_____

_____

49. Identify *one* cell structure found in a producer in this meadow ecosystem that is *not* found in the carnivores. [1]

_____

50. Individuals of some species, such as earthworms, have both male and female sex organs. In many cases, however, these individuals do not fertilize their own eggs.

State *one* genetic advantage of an earthworm mating with another earthworm for the production of offspring. [1]

_____

_____

Base your answers to questions 51 and 52 on the diagram below and on your knowledge of biology. The diagram represents six insect species.

Species E   Species F

51. A dichotomous key to these six species is shown below. Complete the missing information for sections 5.*a*. and 5.*b*. so that the key is complete for all *six* species. [1]

Dichotomous Key

1. a. has small wings ....go to 2
   b. has large wings...go to 3

2. a. has a single pair of wings..........................Species A
   b. has a double pair of wings ....Species B

3. a. has a double pair of wings ....go to 4
   b. has a single pair of wings...........................Species C

4. a. has spots ....go to 5
   b. does not have spots...................................Species D

5. a. _____..............Species E

   b. _____..............Species F

52. Use the key to identify the drawings of species *A*, *B*, *C*, and *D*. Place the letter of each species on the line located below the drawing of the species. [1]

Species ___   Species E   Species ___   Species F   Species ___   Species ___

Base your answers to questions 53 through 55 on the information below and on your knowledge of biology.

Proteins on the surface of a human cell and on a bird influenza virus are represented in the diagram below.

Human Cell   Bird Influenza Virus

53. In the space below, draw a change in the bird influenza virus that would allow it to infect this human cell. [1]

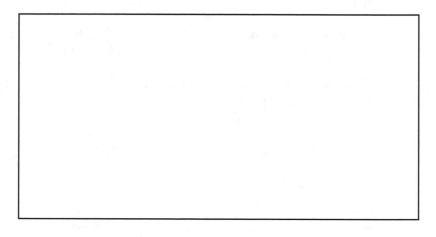

54. Explain how this change in the virus could come about. [1]

_____

_____

55. Identify the relationship that exists between a virus and a human when the virus infects the human. [1]

_____

_____

Part C
Answer all questions in this part. [17]

Directions (56–67): Record your answers in the spaces provided in this examination booklet.

Base your answers to questions 56 and 57 on the information below and on your knowledge of biology.

Insulin is a hormone that has an important role in the maintenance of homeostasis in humans.

56. Identify the structure in the human body that is the usual source of insulin. [1]

_____

57. Identify a substance in the blood, other than insulin, that could change in concentration and indicate a person is not secreting insulin in normal amounts. [1]

_____

Base your answers to questions 58 and 59 on the information below and on your knowledge of biology.

The hedgehog, a small mammal native to Africa and Europe, has been introduced to the United States as an exotic pet species. Scientists have found that hedgehogs can transfer pathogens to humans and domestic animals. Foot-and-mouth viruses, *Salmonella*, and certain fungi are known pathogens carried by hedgehogs. As more and more of these exotic animals are brought into this country, the risk of infection increases in the human population.

58. State *one negative* effect of importing exotic species to the United States. [1]

_____

_____

59. State *one* way the human immune system might respond to an invading pathogen associated with handling a hedgehog. [1]

_____

_____

Base your answers to questions 60 through 62 on the information below and on your knowledge of biology.

The last known wolf native to the Adirondack Mountains of New York State was killed over a century ago. Several environmental groups have recently proposed reintroducing the wolf to the Adirondacks. These groups claim there is sufficient prey to support a wolf population in this area. These prey include beaver, deer, and moose. Opponents of this proposal state that the Adirondacks already have a dominant predator, the Eastern coyote.

60. State *one* effect the reintroduction of the wolf may have on the coyote population within the Adirondacks. Explain why it would have this effect. [1]

_____

_____

_____

61. Explain why the coyote is considered a limiting factor in the Adirondack Mountains. [1]

_____

_____

_____

62. State *one* ecological reason why some individuals might support the reintroduction of wolves to the Adirondacks. [1]

_____

_____

_____

63. You have been assigned to design an experiment to determine the effects of light on the growth of tomato plants. In your experimental design be sure to:
    • state *one* hypothesis to be tested [1]
    • identify the independent variable in the experiment [1]
    • describe the type of data to be collected [1]

_____

_____

64. In some land plants, guard cells are found only on the lower surfaces of the leaves. In some water plants, guard cells are found only on the upper surfaces of the leaves. Explain how guard cells in both land and water plants help maintain homeostasis. In your answer be sure to:
- identify *one* function regulated by the guard cells in leaves [1]
- explain how guard cells carry out this function [1]
- give *one* possible evolutionary advantage of the position of the guard cells on the leaves of land plants [1]

Base your answers to questions 65 and 66 on the information below and on your knowledge of biology.

Scientists are increasingly concerned about the possible effects of damage to the ozone layer.

65. Damage to the ozone layer has resulted in mutations in skin cells that lead to cancer. Will the mutations that caused the skin cancers be passed on to offspring? Support your answer. [1]

_____

_____

_____

_____

66. State *two* specific ways in which an ocean ecosystem will change (other than fewer photosynthetic organisms) if populations of photosynthetic organisms die off as a result of damage to the ozone layer. [2]

_____

_____

_____

_____

_____

67. Lawn wastes, such as grass clippings and leaves, were once collected with household trash and dumped into landfills. Identify *one* way that this practice was harmful to the environment. [1]

_____

_____

_____

_____

Part D

Answer all questions in this part. [13]

Directions (68–80): For those questions that are followed by four choices, write in the space provided the *number* of the choice that, of those given, best completes the statement or answers the question. For all other questions in this part, follow the directions given in the question and record your answers in the spaces provided.

June 2008

68. In preparation for an electrophoresis procedure, enzymes are added to DNA in order to
(1) convert the DNA into gel
(2) cut the DNA into fragments
(3) change the color of the DNA
(4) produce longer sections of DNA   68 ____

69. Paper chromatography is a laboratory technique that is used to
(1) separate different molecules from one another
(2) stain cell organelles
(3) indicate the pH of a substance
(4) compare relative cell sizes   69 ____

70. A marathon runner frequently experiences muscle cramps while running. If he stops running and rests, the cramps eventually go away. The cramping in the muscles most likely results from
(1) lack of adequate oxygen supply to the muscle
(2) the runner running too slowly
(3) the runner warming up before running
(4) increased glucose production in the muscle   70 ____

June 2008   **41**
Living Environment

Base your answers to questions 71 through 73 on the information below and on your knowledge of biology.

A series of investigations was performed on four different plant species. The results of these investigations are recorded in the data table below.

Characteristics of Four Plant Species

Plant Species	Seeds	Leaves	Pattern of Vascular Bundles (structures in stem)	Type of Chlorophyll Present
A	round/small	needle-like	scattered bundles	chlorophyll a and b
B	long/pointed	needle-like	circular bundles	chlorophyll a and c
C	round/small	needle-like	scattered bundles	chlorophyll a and b
D	round/small	needle-like	scattered bundles	chlorophyll b

71. Based on these data, which *two* plant species appear to be most closely related? Support your answer. [1]

Plant species _____ and _____

_____

_____

_____

72. What additional information could be gathered to support your answer to question 71? [1]

_____

_____

_____

73. State *one* reason why scientists might want to know if two plant species are closely related. [1]

_____

_____

_____

Base your answers to questions 74 and 75 on the accompanying data table and on your knowledge of biology.

Dietary Preferences of Finches

Species of Finch	Preferred Foods
A	nuts and seeds
B	worms and insects
C	fruits and seeds
D	insects and seeds
E	nuts and seeds

74. Based on its preferred food, species *B* would be classified as a
(1) decomposer
(2) producer
(3) carnivore
(4) parasite

74 _____

75. Which two species would most likely be able to live in the same habitat without competing with each other for food?
(1) *A* and *C*   (2) *B* and *C*   (3) *B* and *D*   (4) *C* and *E*

75 _____

Base your answers to questions 76 and 77 on the experimental setup shown to the right.

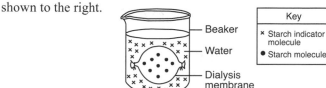

76. On the accompanying diagram, draw in the expected locations of the molecules after a period of one hour. [1]

77. When starch indicator is used, what observation would indicate the presence of starch? [1]

_____

_____

_____

78. State *one* reason why some molecules can pass through a certain membrane, but other molecules can *not*. [1]

_____

_____

_____

79. A plant cell in a microscopic field of view is represented to the right. The width (*w*) of this plant cell is closest to

(1) 200 µm
(2) 800 µm
(3) 1200 µm
(4) 1600 µm

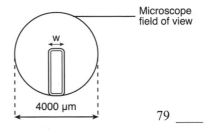

Microscope field of view

w

4000 µm

79 _____

80. The diagram below represents a plant cell in tap water as seen with a compound light microscope.

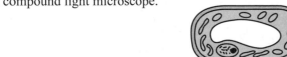

Which diagram best represents the appearance of the cell after it has been placed in a 15% salt solution for two minutes?

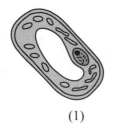   

(1)   (2)   (3)   (4)

80 _____

Answer all questions in this part. [30]

Directions (1–30): For *each* statement or question, write in the space provided the *number* of the word or expression that, of those given, best completes the statement or answers the question.

1. Which statement best describes one of the stages represented in the accompanying diagram?
(1) The mature forest will most likely be stable over a long period of time.
(2) If all the weeds and grasses are destroyed, the number of carnivores will increase.
(3) As the population of the shrubs increases, it will be held in check by the mature forest community.
(4) The young forest community will invade and take over the mature forest community.

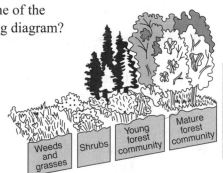

1 _____

2. Which organ system in humans is most directly involved in the transport of oxygen?
(1) digestive   (2) nervous   (3) excretory   (4) circulatory

2 _____

3. Which cell structure contains information needed for protein synthesis?

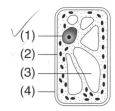

3 _____

4. The human liver contains many specialized cells that secrete bile. Only these cells produce bile because
(1) different cells use different parts of the genetic information they contain
(2) cells can eliminate the genetic codes that they do not need
(3) all other cells in the body lack the genes needed for the production of bile
(4) these cells mutated during embryonic development

4 _____

5. Although identical twins inherit exact copies of the same genes, the twins may look and act differently from each other because
(1) a mutation took place in the gametes that produced the twins
✓(2) the expression of genes may be modified by environmental factors
(3) the expression of genes may be different in males and females
(4) a mutation took place in the zygote that produced the twins   5 _____

6. Which hormone does *not* directly regulate human reproductive cycles?
(1) testosterone   (2) estrogen   (3) insulin   (4) progesterone   6 _____

7. Owls periodically expel a mass of undigested material known as a pellet. A student obtained several owl pellets from the same location and examined the animal remains in the pellets. He then recorded the number of different prey animal remains in the pellets. The student was most likely studying the
(1) evolution of the owl
(2) social structure of the local owl population
✓(3) role of the owl in the local ecosystem
(4) life cycle of the owl   7 _____

8. Which sequence best represents the relationship between DNA and the traits of an organism?

DNA base sequence	DNA base sequence	DNA base sequence	DNA base sequence
Amino acid sequence	Protein shape	Amino acid sequence	Protein function
Protein shape	Amino acid sequence	Protein function	Amino acid sequence
Protein function	Protein function	Protein shape	Protein shape
Trait	Trait	Trait	Trait
✓ (1)	(2)	(3)	(4)

8 _____

9. A sequence of events associated with ecosystem stability is represented below.
sexual reproduction → genetic variation → biodiversity → ecosystem stability
The arrows in this sequence should be read as
✓(1) leads to   (2) reduces   (3) prevents   (4) simplifies   9 _____

10. In some people, the lack of a particular enzyme causes a disease. Scientists are attempting to use bacteria to produce this enzyme for the treatment of people with the disease. Which row in the chart below best describes the sequence of steps the scientists would most likely follow?

Row	Step A	Step B	Step C	Step D
(1)	identify the gene	insert the gene into a bacterium	remove the gene	extract the enzyme
(2)	insert the gene into a bacterium	identify the gene	remove the gene	extract the enzyme
(3)	identify the gene	remove the gene	insert the gene into a bacterium	extract the enzyme
(4)	remove the gene	extract the enzyme	identify the gene	insert the gene into a bacterium

10 _____

11. What will most likely occur as a result of changes in the frequency of a gene in a particular population?
(1) ecological succession
(3) global warming
(2) biological evolution
(4) resource depletion

11 _____

12. The puppies shown in the accompanying photograph are all from the same litter. The differences seen within this group of puppies are most likely due to
(1) overproduction and selective breeding
(2) mutations and elimination of genes
(3) evolution and asexual reproduction
(4) sorting and recombination of genes

12 _____

13. Carbon dioxide makes up less than 1 percent of Earth's atmosphere, and oxygen makes up about 20 percent. These percentages are maintained most directly by
(1) respiration and photosynthesis   (3) synthesis and digestion
(2) the ozone shield   (4) energy recycling in ecosystems

13 _____

14. Which sequence represents the order of some events in human development?
(1) zygote → sperm → tissues → egg
(2) fetus → tissues → zygote → egg
(3) zygote → tissues → organs → fetus
(4) sperm → zygote → organs → tissues

14 _____

15. A variety of plant produces small white fruit. A stem was removed from this organism and planted in a garden. If this stem grows into a new plant, it would most likely produce
(1) large red fruit, only
(2) large pink fruit, only
(3) small white fruit, only
(4) small red and small white fruit on the same plant   15 ____

16. A mutation that can be inherited by offspring would result from
(1) random breakage of chromosomes in the nucleus of liver cells
(2) a base substitution in gametes during meiosis
(3) abnormal lung cells produced by toxins in smoke
(4) ultraviolet radiation damage to skin cells   16 ____

17. The accompanying diagram represents a process that occurs in organisms. Which row in the chart indicates what A and B in the boxes could represent?

17 ____

A	Broken down to	B

Row	A	B
(1)	starch	proteins
(2)	starch	amino acids
(3)	protein	amino acids
(4)	protein	simple sugars

18. Some organs of the human body are represented in the accompanying diagram. Which statement best describes the functions of these organs?
(1) B pumps blood to A for gas exchange.
(2) A and B both produce carbon dioxide, which provides nutrients for other body parts.
(3) A releases antibodies in response to an infection in B.
(4) The removal of wastes from both A and B involves the use of energy from ATP.

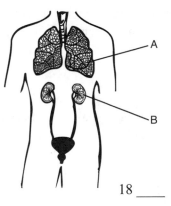

18 ____

19. *Salmonella* bacteria can cause humans to have stomach cramps, vomiting, diarrhea, and fever. The effect these bacteria have on humans indicates that *Salmonella* bacteria are
(1) predators   (3) parasitic fungi
(2) pathogenic organisms   (4) decomposers   19 ____

20. The virus that causes AIDS is damaging to the body because it
(1) targets cells that fight invading microbes
(2) attacks specific red blood cells
(3) causes an abnormally high insulin level
(4) prevents the normal transmission of nerve impulses   20 ____

21. In the leaf of a plant, guard cells help to
(1) block harmful ultraviolet rays that can disrupt chlorophyll production
(2) regulate oxygen and carbon dioxide levels
(3) destroy atmospheric pollutants when they enter the plant
(4) transport excess glucose to the roots   21 ____

22. An antibiotic is effective in killing 95% of a population of bacteria that reproduce by the process shown to the right.
Which statement best describes future generations of these bacteria?
(1) They will be produced by asexual reproduction and will be more resistant to the antibiotic.
(2) They will be produced by sexual reproduction and will be more resistant to the antibiotic.
(3) They will be produced by asexual reproduction and will be just as susceptible to the antibiotic.
(4) They will be produced by sexual reproduction and will be just as susceptible to the antibiotic.   22 ____

23. The size of plant populations can be influenced by the
(1) molecular structure of available oxygen
(2) size of the cells of decomposers
(3) number of chemical bonds in a glucose molecule
(4) type of minerals present in the soil   23 ____

24. Competition between two species occurs when
(1) mold grows on a tree that has fallen in the forest
(2) chipmunks and squirrels eat sunflower seeds in a garden
(3) a crow feeds on the remains of a rabbit killed on the road
(4) a lion stalks, kills, and eats an antelope   24 ____

25. A food chain is illustrated.
The arrows represented as
⟿ most likely indicate

Seaweed → Small fish → Large fish → Shark

(1) energy released into the environment as heat
(2) oxygen produced by respiration
(3) the absorption of energy that has been synthesized
(4) the transport of glucose away from the organism   25 ____

26. If several species of carnivores are removed from an ecosystem, the most likely effect on the ecosystem will be
(1) an increase in the kinds of autotrophs
(2) a decrease in the number of abiotic factors
(3) a decrease in stability among populations
(4) an increase in the rate of succession   26 ____

27. Some people make compost piles consisting of weeds and other plant materials. When the compost has decomposed, it can be used as fertilizer. The production and use of compost is an example of
(1) the introduction of natural predators
(2) the use of fossil fuels
(3) the deforestation of an area
(4) the recycling of nutrients   27 ____

28. Which statement best describes a chromosome?
(1) It is a gene that has thousands of different forms.
(2) It has genetic information contained in DNA.
(3) It is a reproductive cell that influences more than one trait.
(4) It contains hundreds of genetically identical DNA molecules.   28 ____

29. The accompanying graph shows how the level of carbon dioxide in the atmosphere has changed over the last 150,000 years. Which environmental factor has been most recently affected by these changes in carbon dioxide level?

Carbon Dioxide Level

Thousands of Years

(1) light intensity
(2) types of decomposers
(3) size of consumers
(4) atmospheric temperature   29 ____

30. One reason why people should be aware of the impact of their actions on the environment is that
(1) ecosystems are never able to recover once they have been adversely affected
(2) the depletion of finite resources cannot be reversed
(3) there is a decreased need for new technology
(4) there is a decreased need for substances produced by natural processes   30 ____

Part B–1
Answer all questions in this part. [11]
Directions (31–41): For *each* statement or question, write in the space provided the *number* of the word or expression that, of those given, best completes the statement or answers the question.

31. The accompanying diagram represents the process used in 1996 to clone the first mammal, a sheep named Dolly.

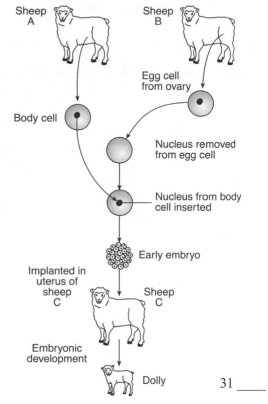

Which statement concerning Dolly is correct?
(1) Gametes from sheep *A* and sheep *B* were united to produce Dolly.
(2) The chromosome makeup of Dolly is identical to that of sheep *A*.
(3) Both Dolly and sheep *C* have identical DNA.
(4) Dolly contains genes from sheep *B* and sheep *C*.

31 ____

32. The accompanying diagram represents a cell. Which statement concerning ATP and activity within the cell is correct?
(1) The absorption of ATP occurs at structure *A*.
(2) The synthesis of ATP occurs within structure *B*.
(3) ATP is produced most efficiently by structure *C*.
(4) The template for ATP is found in structure *D*.

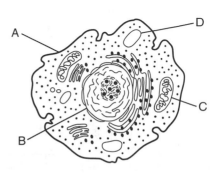

32 ____

33. The accompanying diagram illustrates some functions of the pituitary gland. The pituitary gland secretes substances that, in turn, cause other glands to secrete different substances. Which statement best describes events shown in the diagram?

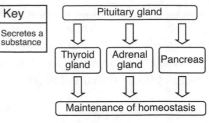

(1) Secretions help the body to respond to changes from the normal state.
(2) The raw materials for the synthesis of secretions come from nitrogen.
(3) The secretions of all glands speed blood circulation in the body.
(4) Secretions provide the energy needed for metabolism.   33 ____

34. A pond ecosystem is shown in the accompanying diagram. Which statement describes an interaction that helps maintain the dynamic equilibrium of this ecosystem?
(1) The frogs make energy available to this ecosystem through the process of photosynthesis.
(2) The algae directly provide food for both the rotifers and the catfish.
(3) The green-backed heron provides energy for the mosquito larvae.
(4) The catfish population helps control the populations of water boatman and water fleas.

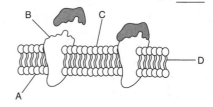

(Not drawn to scale)   34 ____

35. The accompanying diagram represents a portion of a cell membrane. Which structure may function in the recognition of chemical signals?

(1) A   (2) B   (3) C   (4) D   35 ____

36. Which species in the accompanying chart is most likely to have the fastest rate of evolution?

Species	Reproductive Rate	Environment
A	slow	stable
B	slow	changing
C	fast	stable
D	fast	changing

(1) A   (3) C
(2) B   (4) D   36 ____

37. Some evolutionary pathways are represented in the accompanying diagram. An inference that can be made from information in the diagram is that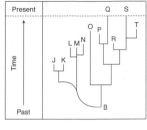
(1) many of the descendants of organism *B* became extinct
(2) organism *B* was probably much larger than any of the other organisms represented
(3) most of the descendants of organism *B* successfully adapted to their environment and have survived to the present time
(4) the letters above organism *B* represent members of a single large population with much biodiversity   37 _____

Base your answers to questions 38 and 39 on the accompanying diagram that represents an energy pyramid in a meadow ecosystem and on your knowledge of biology.

38. Which species would have the largest amount of available energy in this ecosystem?
(1) *A*   (3) *C*
(2) *B*   (4) *E*   38 _____

39. Which two organisms are carnivores?
(1) *A* and *B*   (2) *A* and *E*   (3) *B* and *D*   (4) *C* and *E*   39 _____

40. The kit fox and red fox species are closely related. The kit fox lives in the desert, while the red fox inhabits forests. Ear size and fur color are two differences that can be observed between the species. An illustration of these two species is shown. Which statement best explains how the differences between these two species came about?

Red Fox

(1) Different adaptations developed because the kit fox preferred hotter environments than the red fox.
(2) As the foxes adapted to different environments, differences in appearance evolved.
(3) The foxes evolved differently to prevent overpopulation of the forest habitat.
(4) The foxes evolved differently because their ancestors were trying to avoid competition.   40 _____

Kit Fox

41. An ecosystem is represented below.

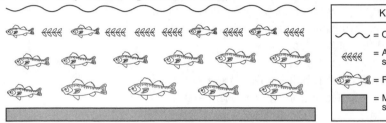

The organisms represented as 𝘴𝘴𝘴𝘴 are found in the area shown due to which factor?

(1) pH   (2) sediment   (3) light intensity   (4) colder temperature   41 ____

Part B–2
Answer all questions in this part. [14]

Directions (42–51): For those questions that are followed by four choices, record in the space provided the *number* preceding the choice that, of those given, best completes the statement or answers the question. For all other questions in this part, follow the directions given in the question and record your answers in the spaces provided.

42. The graphs below show dissolved oxygen content, sewage waste content, and fish populations in a lake between 1950 and 1970.

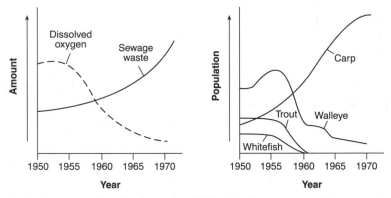

State what happened to the amount of dissolved oxygen and the number of fish species as the amount of sewage waste increased. [1]

_____

_____

_____

Base your answers to questions 43 through 46 on the information below and on your knowledge of biology.

$$C_6H_{12}O_6 \rightarrow 2C_2H_5OH + 2CO_2$$

glucose   ethyl   carbon
       alcohol   dioxide

Yeast cells carry out the process of cellular respiration as shown in the equation.

An investigation was carried out to determine the effect of temperature on the rate of cellular respiration in yeast. Five experimental groups, each containing five fermentation tubes, were set up. The fermentation tubes all contained the same amounts of water, glucose, and yeast. Each group of five tubes was placed in a water bath at a different temperature. After 30 minutes, the amount of gas produced (*D*) in each fermentation tube was measured in milliliters. The average for each group was calculated. A sample setup and the data collected are shown.

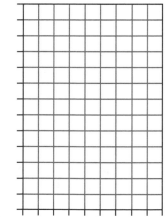

Average Amount of Gas Produced (D)
After 30 Minutes at Various Temperatures

Group	Temperature (°C)	D (mL)
1	5	0
2	20	5
3	40	12
4	60	6
5	80	3

Directions (43 and 44): Using the information in the data table, construct a line graph on the grid below, following the directions below.

Average Amount of Gas Produced at Various Temperatures

43. Mark an appropriate scale on each labeled axis. [1]

44. Plot the data from the data table. Surround each point with a small circle, and connect the points. [1] Example: ⊙—⊙

45. The maximum rate of cellular respiration in yeast occurred at which temperature?
(1) 5°C   (3) 40°C
(2) 20°C   (4) 60°C   45 _____

Average Amount of Gas Produced (mL)

Temperature (°C)

46. Compared to the other tubes at the end of 30 minutes, the tubes in group 3 contained the
(1) smallest amount of CO_2
(2) smallest amount of glucose
(3) smallest amount of ethyl alcohol
(4) same amounts of glucose, ethyl alcohol, and CO_2 46 _____

Base your answers to questions 47 through 49 on the information below and on your knowledge of biology.

An ecologist made some observations in a forest ecosystem over a period of several days. Some of the data collected are shown in the table below.

Observations in a Forest Environment

Date	Observed Feeding Relationships	Ecosystem Observations
6/2	• white-tailed deer feeding on maple tree leaves • woodpecker feeding on insects • salamander feeding on insects	• 2 cm of rain in 24 hours
6/5	• fungus growing on a maple tree • insects feeding on oak trees	• several types of sedimentary rock are in the forest
6/8	• woodpecker feeding on insects • red-tailed hawk feeding on chipmunk	• air contains 20.9% oxygen
6/11	• chipmunk feeding on insects • insect feeding on maple tree leaves • chipmunk feeding on a small salamander	• soil contains phosphorous

47. On the diagram below, complete the food web by placing the names of *all* the organisms in the correct locations. [1]

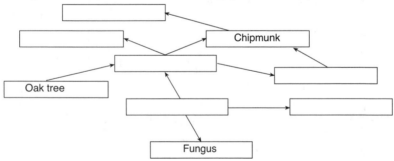

48. Identify *one* producer recorded by the ecologist in the data table. [1] _____

49. Which statement describes how one biotic factor of the forest uses one of the abiotic factors listed in the data table?

(1) Trees absorb water as a raw material for photosynthesis.
(2) Insects eat and digest the leaves of trees.
(3) Erosion of sedimentary rock adds phosphorous to the soil.
(4) Fungi release oxygen from the trees back into the air.   49 ____

50. Fill in all of the blanks in parts 2 and 3 of the dichotomous key below, so that it contains information that could be used to identify the four animals shown below. [2]

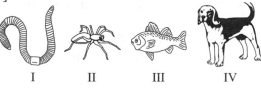

I   II   III   IV

Dichotomous Key

1.   a. Legs present..Go to 2
     b. Legs not present..Go to 3

Characteristic		**Organism**
2. a. _____	....	_____
b. _____	....	_____
3. a. _____	....	_____
b. _____	....	_____

51. The human female reproductive system is represented in the accompanying diagram.

Complete boxes 1 through 4 in the chart below using the information from the diagram. [4]

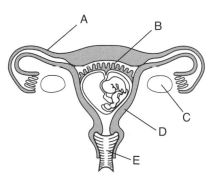

Name of Structure	Letter on Diagram	Function of Structure
1 _____	2 _____	produces gametes
uterus	D	3 _____
4 _____	B	transports oxygen directly to the embryo

Part C
Answer all questions in this part. [17]

Directions (52–59): **Record your answers in the spaces provided.**

52. Humans have many interactions with the environment. Briefly describe how human activities can affect the environment of organisms living 50 years from now.

In your answer, be sure to:
- identify *one* human activity that could release chemicals harmful to the environment [1]
- identify the chemical released by the activity [1]
- state *one* effect the release of this chemical would most likely have on future ecosystems [1]
- state *one* way in which humans can reduce the production of this chemical to lessen its effect on future ecosystems [1]

_____

_____

_____

_____

_____

_____

_____

_____

_____

_____

_____

_____

_____

_____

_____

_____

_____

_____

_____

53. Plants respond to their environment in many different ways. Design an experiment to test the effects of *one* environmental factor, chosen from the list below, on plant growth.

Acidity of precipitation   Temperature   Amount of water

In your answer, be sure to:
- identify the environmental factor you chose
- state *one* hypothesis the experiment would test [1]
- state how the control group would be treated differently from the experimental group [1]
- state *two* factors that must be kept the same in both the experimental and control groups [1]
- identify the independent variable in the experiment [1]
- label the columns on the data table below for the collection of data in your experiment [1]

Environmental factor: _____

_____

_____

_____

_____

_____

_____

_____

_____

_____

_____

_____

_____

_____

Data Table

Base your answer to question 54 on the article below and on your knowledge of biology.

Power plan calls for windmills off beach

The Associated Press

Several dozen windmills taller than the Statue of Liberty will crop up off Long Island — the first source of off-shore wind power outside of Europe, officials said.

The Long Island Power Authority [LIPA] expects to choose a company to build and operate between 35 and 40 windmills in the Atlantic Ocean off Jones Beach, The New York Times reported Sunday [May 2, 2004]. Cost and completion date are unknown.

Energy generated by the windmills would constitute about 2 percent of LIPA's total power use. They are expected to produce 100 to 140 megawatts, enough to power 30,000 homes....

But some Long Island residents oppose the windmills, which they fear will create noise, interfere with fishing, and mar ocean views....

Source: "Democrat and Chronicle", Rochester, NY 5/3/04

54. State *two* ways that the use of windmills to produce energy would be beneficial to the environment. [2]

(1) _____

(2) _____

55. Importing a foreign species, either intentionally or by accident, can alter the balance of an ecosystem. State *one* specific example of an imported species that has altered the balance of an ecosystem and explain how it has disrupted the balance in that ecosystem. [2]

_____

_____

_____

_____

_____

Base your answers to questions 56 through 59 on the passage below and on your knowledge of biology.

Avian (Bird) Flu

Avian flu virus H5N1 has been a major concern recently. Most humans have not been exposed to this strain of the virus, so they have not produced the necessary protective substances. A vaccine has been developed and is being made in large quantities. However, much more time is needed to manufacture enough vaccine to protect most of the human population of the world.

Most flu virus strains affect the upper respiratory tract, resulting in a runny nose and sore throat. However, the H5N1 virus seems to go deeper into the lungs and causes severe pneumonia, which may be fatal for people infected by this virus.

So far, this virus has not been known to spread directly from one human to another. As long as H5N1 does not change to another strain that can be transferred from one human to another, a worldwide epidemic of the virus probably will not occur.

56. State *one* difference between the effect on the human body of the usual forms of flu virus and the effect of H5N1. [1]

_____

_____

_____

57. Identify the type of substance produced by the human body that protects against antigens, such as the flu virus. [1]

_____

58. State what is in a vaccine that makes the vaccine effective. [1]

_____

59. Identify *one* event that could result in the virus changing to a form able to spread from human to human. [1]

_____

Part D
Answer all questions in this part. [13]

Directions (60–72): For those questions that are followed by four choices, record in the space provided the *number* of the choice, that, of those given, best completes the statement or answers the question. For all other questions in this part, follow the directions given in the question and record your answers in the spaces provided.

Amino Acid Differences

60. The accompanying data table shows the number of amino acid differences in the hemoglobin molecules of several species compared with amino acids in the hemoglobin of humans.

Species	Number of Amino Acid Differences
human	0
frog	67
pig	10
gorilla	1
horse	26

Based on the information in the data table, write the names of the organisms from the table in their correct positions on the evolutionary tree below. [1]

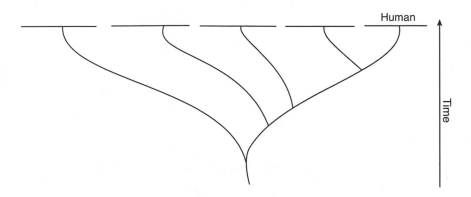

61. Explain why comparing the vein patterns of several leaves is a less reliable means of determining the evolutionary relationship between two plants than using gel electrophoresis.[1]

_____

_____

_____

Base your answer to question 62 on the information and diagram below and on your knowledge of biology.

An enzyme and soluble starch were added to a test tube of water and kept at room temperature for 24 hours. Then, 10 drops of glucose indicator solution were added to the test tube, and the test tube was heated in a hot water bath for 2 minutes.

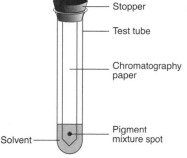

Water containing an enzyme and soluble starch — 24 hours → 10 drops glucose indicator solution added
Water bath
Hot plate

62. The test was performed in order to
(1) measure the quantity of fat that is converted to starch
(2) determine if digestion took place
(3) evaporate the water from the test tube
(4) cause the enzyme to bond to the water   62 _____

63. A chromatography setup is shown to the right. Identify *one* error in the setup. [1]

Stopper
Test tube
Chromatography paper
Solvent
Pigment mixture spot

_____

_____

_____

_____

_____

_____

Base your answers to questions 64 through 66 on the information and data table below and on your knowledge of biology.

During a laboratory activity, a group of students obtained the data to the right.

Pulse Rate Before and After Exercise

Student Tested	Pulse Rate at Rest (beats/min)	Pulse Rate After Exercise (beats/min)
A	70	97
B	74	106
C	83	120
D	60	91
E	78	122
Group Average		107

64. Which procedure would increase the validity of the conclusions drawn from the results of this experiment?
(1) increasing the number of times the activity is repeated
(2) changing the temperature in the room
(3) decreasing the number of students participating in the activity
(4) eliminating the rest period before the resting pulse rate is taken 64 _____

65. Calculate the group average for the resting pulse rate. [1] _____ **beats/min**

66. A change in pulse rate is related to other changes in the body. Write the name of *one* organ that is affected when a person runs a mile and describe *one* change that occurs in this organ. [1]

Organ: _____

_____

_____

Base your answers to questions 67 through 69 on the information and diagram below and on your knowledge of biology.

A wet mount of red onion cells as seen with a compound light microscope is shown to the right.

67. Which diagram best illustrates the technique that would most likely be used to add salt to these cells?

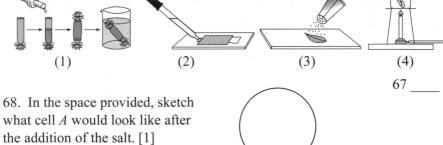

(1)	(2)	(3)	(4)

67 _____

68. In the space provided, sketch what cell *A* would look like after the addition of the salt. [1]

69. Which substance would most likely be used to return the cells to their original condition?
(1) starch indicator   (3) glucose indicator solution
(2) dialysis tubing   (4) distilled water   69 _____

70. DNA electrophoresis is used to study evolutionary relationships of species. The accompanying diagram shows the results of DNA electrophoresis for four different animal species. Which species has the most DNA in common with species *A*?
(1) *X* and *Y*, only (3) *Z*, only
(2) *Y*, only   (4) *X*, *Y*, and *Z*

70 _____

Species A	Species X	Species Y	Species Z
—	—	—	—
		—	
			—
—	—	—	
—	—	—	—
	—	—	=
—		—	
	—		—

Base your answers to questions 71 and 72 on the diagram below that shows variations in the beaks of finches in the Galapagos Islands and on your knowledge of biology.

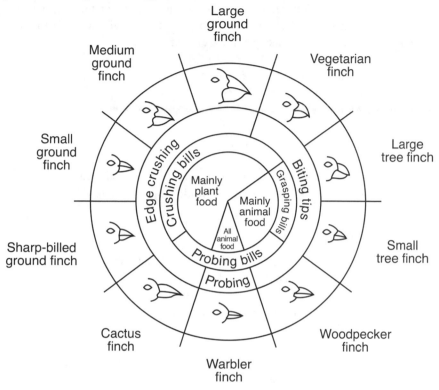

From: *Galapagos: A Natural History Guide*

71. The diversity of species seen on the Galapagos Islands is mostly due to
(1) gene manipulation by scientists
(2) gene changes resulting from mitotic cell division
(3) natural selection
(4) selective breeding   71 ____

72. State *one* reason why large ground finches and large tree finches can coexist on the same island. [1]

_____

_____

_____

Answer all questions in this part. [30]

Directions (1–30): For *each* statement or question, write in the space provided the *number* of the word or expression that, of those given, best completes the statement or answers the question.

1. Why is a mushroom considered a heterotroph?
(1) It manufactures its own food.
(2) It divides by mitosis.
(3) It transforms light energy into chemical energy.
(4) It obtains nutrients from its environment.   1 _____

2. Three days after an organism eats some meat, many of the organic molecules originally contained in the meat would be found in newly formed molecules of
(1) glucose   (2) protein   (3) starch   (4) oxygen   2 _____

3. Which body system is correctly paired with its function?
(1) excretory—produces antibodies to fight disease-causing organisms
(2) digestive—produces hormones for storage and insulation
(3) circulatory—transports materials for energy release in body cells
(4) respiratory—collects waste material for digestion   3 _____

4. Which statement best explains why some cells in the reproductive system only respond to certain hormones?
(1) These cells have different DNA than the cells in other body systems.
(2) These cells have specific types of receptors on their membranes.
(3) Reproductive system cells could be harmed if they made contact with hormones from other body systems.
(4) Cells associated with the female reproductive system only respond to the hormone testosterone.   4 _____

5. In the accompanying cell shown, which lettered structure is responsible for the excretion of most cellular wastes?
(1) *A*   (3) *C*
(2) *B*   (4) *D*

(A)—
(B)—
(C)—
(D)—   5 _____

6. What is the main function of a vacuole in a cell?
(1) storage (3) synthesis of molecules
(2) coordination (4) release of energy 6 _____

7. If 15% of a DNA sample is made up of thymine, T, what percentage of the sample is made up of cytosine, C?
(1) 15% (2) 35% (3) 70% (4) 85% 7 _____

8. Global warming has been linked to a *decrease* in the
(1) size of the polar ice caps (3) rate of species extinction
(2) temperature of Earth (4) rate of carbon dioxide production 8 _____

9. Several structures are labeled in the accompanying diagram of a puppy. Every cell in each of these structures contains 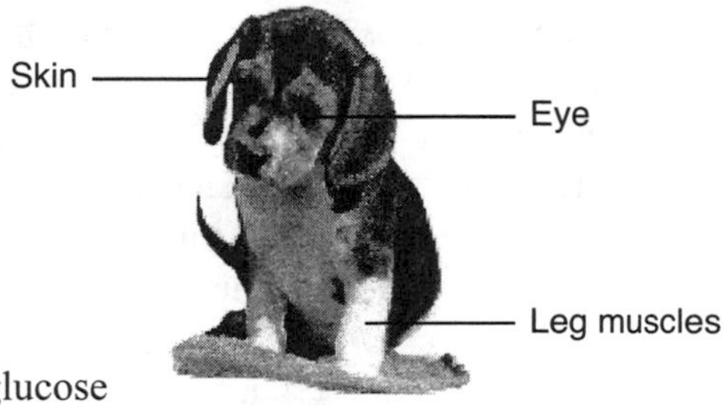
(1) equal amounts of ATP
(2) identical genetic information
(3) proteins that are all identical
(4) organelles for the synthesis of glucose

Skin

Eye

Leg muscles

9 _____

10. A characteristic that an organism exhibits during its lifetime will only affect the evolution of its species if the characteristic
(1) results from isolation of the organism from the rest of the population
(2) is due to a genetic code that is present in the gametes of the organism
(3) decreases the number of genes in the body cells of the organism
(4) causes a change in the environment surrounding the organism 10 _____

11. Agriculturists have developed some varieties of vegetables from common wild mustard plants, which reproduce sexually. Which statement best explains the development of these different varieties of vegetables?
(1) Different varieties can develop from a single species as a result of the recombination of genetic information.
(2) Different species can develop from a single species as a result of the effect of similar environmental conditions.
(3) Mutations will occur in the genes of a species only if the environment changes.
(4) Variations in a species will increase when the rate of mitosis is decreased. 11 _____

12. The accompanying diagram represents a technique used in some molecular biology laboratories. This technique is a type of

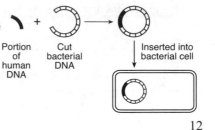

Portion of human DNA
Cut bacterial DNA
Inserted into bacterial cell

(1) chromatography
(2) gel electrophoresis
(3) direct harvesting
(4) genetic engineering

12 _____

13. A species of bird known as Bird of Paradise has been observed in the jungles of New Guinea. The males shake their bodies and sometimes hang upside down to show off their bright colors and long feathers to attract females. Females usually mate with the "flashiest" males. These observations can be used to support the concept that
(1) unusual courtship behaviors lead to extinction
(2) some organisms are better adapted for asexual reproduction
(3) homeostasis in an organism is influenced by physical characteristics
(4) behaviors that lead to reproductive success have evolved

13 _____

14. Which statement concerning the evolution of species A, B, C, D, and E is supported by the accompanying diagram?
(1) Species B and C can be found in today's environments.
(2) Species A and D evolved from E.
(3) Species A and C can still interbreed.
(4) Species A, B, and E all evolved from a common ancestor and all are successful today.

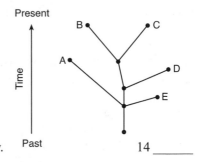

14 _____

15. The diagram below represents a process that occurs during human reproduction.

(Not drawn to scale)

The process represented by the arrow will ensure that the
(1) zygote contains a complete set of genetic information
(2) gametes contain a complete set of genetic information
(3) zygote contains half of the genetic information
(4) gametes contain half of the genetic information

15 _____

16. Even though identical twins have the same genetic material, they may develop slightly different characteristics because
(1) each twin receives different chromosomes from the egg
(2) one twin may only have genes from the father
(3) gene expression may be influenced by factors that switch genes on and off
(4) a gene mutation may have occurred before the zygote divided   16 _____

17. What normally happens immediately after fertilization in sexual reproduction?
(1) specialization of cells to form a fetus from an egg
(2) production of daughter cells having twice the number of chromosomes as the parent cell
(3) production of daughter cells having half the number of chromosomes as the parent cell
(4) division of cells resulting in the development of an embryo from a zygote   17 _____

18. The human female reproductive system is represented in the diagram below.

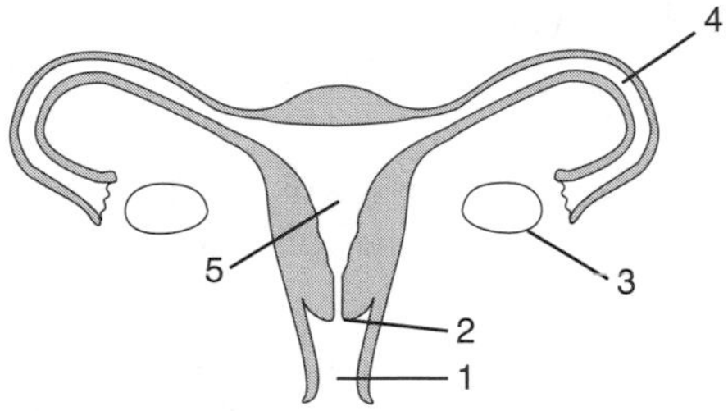

Production of gametes and support of the fetus normally occur in structures
(1) 1 and 2   (2) 2 and 4   (3) 3 and 5   (4) 4 and 5   18 _____

19. Essential materials needed for development are transported to a human fetus through the
(1) reproductive hormones   (3) placenta
(2) egg cell   (4) ovaries   19 _____

20. The failure to regulate the pH of the blood can affect the activity of
(1) enzymes that clot blood
(2) red blood cells that make antibodies
(3) chlorophyll that carries oxygen in the blood
(4) DNA that controls starch digestion in the blood   20 _____

June 2010

21. Young birds that have been raised in isolation from members of their species build nests characteristic of their species. This suggests that the nest-building behavior is
(1) genetically inherited from parents
(2) learned by watching members of their species
(3) a disadvantage to the survival of the species
(4) a direct result of the type of food the bird eats   21 _____

22. Some people with spinal cord injuries do not sweat below the area of the injury. Without the ability to sweat, the human body temperature begins to rise. Which statement would best describe this situation?
(1) Feedback mechanisms regulate blood sugar levels.
(2) Gene mutations are increased.
(3) Energy from ATP is not available.
(4) Dynamic equilibrium is disrupted.   22 _____

23. Decomposers are necessary in an ecosystem because they
(1) produce food for plants by the process of photosynthesis
(2) provide energy for plants by the process of decay
(3) can rapidly reproduce and evolve
(4) make inorganic materials available to plants   23 _____

24. A manatee is a water-dwelling herbivore on the list of endangered species. If manatees were to become extinct, what would be the most likely result in the areas where they had lived?
(1) The biodiversity of these areas would not be affected.
(2) Certain producer organisms would become more abundant in these areas.
(3) Other manatees would move into these areas and restore the population.
(4) Predators in these areas would occupy higher levels on the energy pyramid.   24 _____

25. A serious threat to biodiversity is
(1) habitat destruction   (3) competition within a species
(2) maintenance of food chains   (4) a stable population size   25 _____

26. Which action will result in the greatest *decrease* in rain forest stability?
(1) removing one species of plant for medicine
(2) harvesting nuts from some trees
(3) cutting down all the trees for lumber
(4) powering all homes with wind energy   26 _____

27. One way that humans could have a positive impact on local environments is to
(1) generate waste products as a result of technological advances
(2) use resources that are renewable
(3) increase planting large areas of one crop
(4) increase the use of pesticides   27 _____

28. Which statement provides evidence that evolution is still occurring at the present time?
(1) The extinction rate of species has decreased in the last 50 years.
(2) Many bird species and some butterfly species make annual migrations.
(3) New varieties of plant species appear more frequently in regions undergoing climatic change.
(4) Through cloning, the genetic makeup of organisms can be predicted.   28 _____

29. The diagram below represents the various stages of ecological succession in New York State.

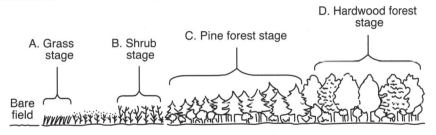

If the ecosystem is not altered, which stage would be the most stable?
(1) grass   (2) shrub   (3) pine forest   (4) hardwood forest   29 _____

30. Because of an attractive tax rebate, a homeowner decides to replace an oil furnace heating system with expensive solar panels. The trade-offs involved in making this decision include
(1) high cost of solar panels, reduced fuel costs, and lower taxes
(2) low cost of solar panels, increased fuel costs, and higher taxes
(3) increased use of fuel, more stable ecosystems, and less availability of solar radiation
(4) more air pollution, increased use of solar energy, and greater production of oil   30 _____

Part B–1

Answer all questions in this part. [13]

Directions (31–43): For *each* statement or question, write in the space provided the *number* of the word or expression that, of those given, best completes the statement or answers the question.

31. A clear plastic ruler is placed across the middle of the field of view of a compound light microscope. A row of cells can be seen under low-power magnification (100×).

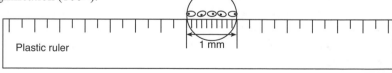

1 mm = 1000 μm

What is the average length of a single cell in micrometers (μm)?
(1) 10 μm   (2) 100 μm   (3) 200 μm   (4) 2000 μm   31 _____

32. The accompanying graph represents the populations of two different species in an ecosystem over a period of several years. Which statement is a possible explanation for the changes shown?
(1) Species *A* is better adapted to this environment.
(2) Species *A* is a predator of species *B*.
(3) Species *B* is better adapted to this environment.
(4) Species *B* is a parasite that has benefited species *A*.

Population Changes in an Ecosystem

Number of Individuals

Species A

Species B

Time (years) ——————→

32 _____

33. A mineral supplement designed to prevent the flu was given to two groups of people during a scientific study. Dosages of the supplement were measured in milligrams per day, as shown in the accompanying table. After 10 weeks, neither group reported a case of the flu. Which procedure would have made the outcome of this study more valid?

Supplement Dosages

Group	Dosage (mg/day)
A	100
B	200

(1) test only one group with 200 mg of the supplement
(2) test the supplement on both groups for 5 weeks instead of 10 weeks
(3) test a third group that receives 150 mg of the supplement
(4) test a third group that does not receive the supplement   33 _____

34. The accompanying diagram shows a normal gene sequence and three mutated sequences of a segment of DNA. Which row in the chart below correctly identifies the cause of each type of mutation?

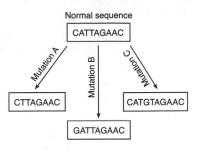

Row	Mutation A	Mutation B	Mutation C
(1)	deletion	substitution	insertion
(2)	insertion	substitution	deletion
(3)	insertion	deletion	substitution
(4)	deletion	insertion	substitution

34 _____

Base your answers to questions 35 and 36 on the accompanying energy pyramid and on your knowledge of biology.

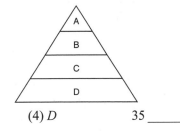

35. Which level includes organisms that receive their energy from level *B*?
(1) *A*   (2) *B*   (3) *C*   (4) *D*   35 _____

36. Which level includes organisms that get their energy exclusively from a source other than the organisms in this ecosystem?
(1) *A*   (2) *B*   (3) *C*   (4) *D*   36 _____

37. The chart compares the size of three structures: a gene, a nucleus, and a chromosome. Based on this information, structure *A* would most likely be a

Size	Structure
smallest in size	A
↓	B
greatest in size	C

(1) chromosome that is part of structure *C*
(2) chromosome that contains structures *B* and *C*
(3) nucleus that contains both structure *B* and structure *A*
(4) gene that is part of structure *B*

37 _____

38. The accompanying diagram shows molecules represented by *X* both outside and inside of a cell. A process that would result in the movement of these molecules out of the cell requires the use of
(1) DNA   (3) antigens
(2) ATP   (4) antibodies

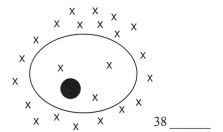

38 _____

39. Which statement most accurately predicts what would happen in the aquarium shown to the right if it were tightly covered and maintained in natural light for one month?
(1) The water temperature would rapidly decrease.
(2) The process of respiration in the snail would decrease.
(3) The rate of reproduction of the fish would be affected.
(4) The organisms would probably survive because materials would cycle.

Microorganisms

39 _____

40. The data table below shows an effect of secondhand smoke on the birth weight of babies born to husbands and wives living together during pregnancy.

Effect of Secondhand Smoke on Birth Weight

	Wife: Nonsmoker Husband: Nonsmoker	Wife: Nonsmoker Husband: Smoker
Number of Couples	837	529
Average Weight of Baby at Birth	3.2 kg	2.9 kg

Based on these data, a reasonable conclusion that can be drawn about secondhand smoke during pregnancy is that secondhand smoke
(1) is unable to pass from the mother to the fetus
(2) slows the growth of the fetus
(3) causes mutations in cells of the ovaries
(4) blocks the receptors on antibody cells

40 _____

41. A limiting factor unique to a field planted with corn year after year is most likely
(1) temperature   (2) sunlight   (3) water   (4) soil nutrients   41 _____

Base your answers to questions 42 and 43 on the information below and on your knowledge of biology.

After the Aswan High Dam was built on the Nile River, the rate of parasitic blood-fluke infection doubled in the human population near the dam. As a result of building the dam, the flow of the Nile changed. This changed the habitat, which resulted in an increase in its population of a certain aquatic snail. The snails, which were infected, released larvae of the fluke. These larvae then infected humans.

42. This situation best illustrates that
(1) the influence of humans on a natural system is always negative in the long term
(2) the influence of humans on a natural system can have unpredictable negative impacts
(3) human alteration of an ecosystem does not need to be studied to avoid ecological disaster
(4) human alteration of an ecosystem will cause pollution and loss of finite resources   42 _____

43. The role of the snail may be described as a
(1) host   (2) parasite   (3) producer   (4) decomposer   43 _____

Part B–2
Answer all questions in this part. [12]
Directions (44– 55): **For questions in this part, follow the directions given in the question and record your answers in the spaces provided.**

44. The accompanying table shows the abundance of some greenhouse gases in the atmosphere. Identify the most abundant greenhouse gas and state *one* human activity that is a source of this gas. [1]

Abundance of Some Atmospheric Greenhouse Gases

Greenhouse Gases	Abundance (%)
carbon dioxide (CO_2)	99.438
methane (CH_4)	0.471
nitrous oxide (N_2O)	0.084
other gases (CFCs, etc.)	0.007
Total	**100.000**

Greenhouse gas: _____

_____

_____

_____

45. The United States government does not allow travelers from foreign countries to bring plants, fruits, vegetables, animals, or other living organisms into this country. State *one* biological reason for keeping these out of the United States. [1]

_____

_____

_____

_____

Base your answers to questions 46 through 49 on the information and data table below and on your knowledge of biology.

Birds colliding with aircraft either on the ground or in the air create problems for the Air Force. An organization known as BASH (Bird Aircraft Strike Hazard) studied the impact of birds colliding with aircraft. In 2001, there were 3854 bird collisions reported at a total cost to the Air Force of over 31 million dollars in damage—approximately eight thousand dollars per collision. August, September, and October were the busiest months with 1442 collisions. Nearly 50% of all these collisions occurred in the airfield environment, an environment that can most easily be controlled.

The top five species of birds involved in these collisions are listed in the accompanying data table.

Top Five Bird Species Involved in Collisions in 2001

Type of Bird	Number of Collisions
American mourning dove (species A)	123
horned lark (species B)	100
barn swallow (species C)	83
American cliff swallow (species D)	55
American robin (species E)	55

Source of data: Bird Aircraft Strike Hazard by Matt Granger, http://www.find.articles.com

Directions (46–47): Using the information in the data table, construct a bar graph on the grid, following the directions below.

46. Mark an appropriate scale on the axis labeled "Number of Collisions." [1]

47. Construct vertical bars to represent the data. Shade in each bar. [1]

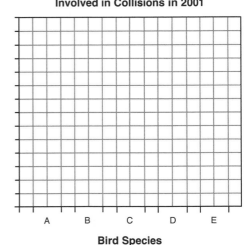

Top Five Bird Species Involved in Collisions in 2001

Number of Collisions

Bird Species

A   B   C   D   E

48. Is the problem with birds and aircraft limited to birds living on or near airport grounds? Support your answer using information from the passage. [1]

_____

_____

49. State *one* possible reason that the greatest number of bird collisions occurs during August, September, and October. [1]

_____

_____

Base your answers to questions 50 through 53 on the information below and on your knowledge of biology.

The Control of Transpiration

Plants normally lose water from openings (stomates) in their leaves. The water loss typically occurs during daylight hours when plants are exposed to the Sun. This water loss, known as transpiration, is both beneficial and harmful to plants.

Scientists believe wind and high temperatures increase the rate of transpiration, but the size of each stomate opening can be regulated. Reducing the size of the openings during drought conditions may help reduce the dehydration and wilting that would otherwise occur.

A leaf may lose more than its own weight in water each day. Transpiration also lowers the internal temperature of the leaf as water evaporates. On hot days, temperatures in the leaves may be from 3° to 15°C cooler than the outside air. With stomates open, vital gases may be exchanged between the leaf tissues and the outside environment.

Researchers have also found many plants that use another response when leaf temperatures rise. Special molecules known as heat shock proteins are produced by plant cells and help to hold enzymes in their functional shapes.

50. State *one* way transpiration is beneficial to plants. [1]

_____

_____

51. Identify *two* of the "vital gases" that are exchanged between leaf tissues and the outside environment. [1]

_____ and _____

52. Identify the specific leaf structures that
regulate the opening and closing of stomates. [1] _____

53. Explain why it is important for plants to "hold enzymes in their functional shapes." [1]

_____

_____

_____

54. The accompanying graph shows the growth of a population of coyotes in a wilderness area. State *one* possible cause for the population decrease at *X*. [1]

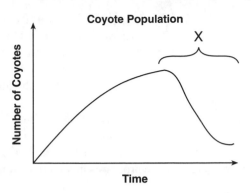

Coyote Population

X

Number of Coyotes

Time

_____

_____

_____

55. The information in the accompanying chart represents the sex chromosome arrangement in humans and birds. Sex chromosomes contain genes involved in sex determination.

Sex Chromosomes in Animals

Animal	Female	Male
humans	XX	XY
birds	ZW	ZZ

In humans, it is the male gamete that is responsible for determining the sex of the offspring. Identify which type of gamete determines the sex of the offspring in birds. Support your answer. [1]

Type of Gamete: _____

_____

_____

_____

Part C

Answer all questions in this part. [17]

Directions (56–64): Record your answers in the spaces provided.

Base your answer to question 56 on the information below and on your knowledge of biology.

Help for Aging Memories

As aging occurs, the ability to form memories begins to decrease. Research has shown that an increase in the production of a certain molecule, BDNF, seems to restore the processes involved in storing memories. BDNF is found in the central nervous system and seems to be important in maintaining nerve cell health. Researchers are testing a new drug that seems to increase the production of BDNF.

56. Design an experiment to test the effectiveness of the new drug to increase the production of BDNF in the brains of rats. In your answer, be sure to:
- state the hypothesis your experiment will test [1]
- describe how the control group will be treated differently from the experimental group [1]
- identify *two* factors that must be kept the same in both the experimental and control groups [1]
- identify the dependent variable in your experiment [1]

_____

_____

_____

_____

_____

_____

_____

_____

_____

_____

_____

_____

_____

_____

_____

_____

_____

_____

Base your answers to questions 57 through 59 on the information below and on your knowledge of biology.

Rabbits eat plants and in turn are eaten by predators such as foxes and wolves. A population of rabbits is found in which a few have a genetic trait that gives them much better than average leg strength.

57. Predict how the frequency of the trait for above average leg strength would be expected to change in the population over time. Explain your prediction. [1]

_____

_____

58. State what is likely to happen to the rabbits in the population that do *not* have the trait for above average leg strength. [1]

_____

_____

59. It was later discovered that the rabbits born with the trait for above average leg strength also inherited the trait for poor eyesight. Taking into account this new information, explain how your predictions would change. Support your answer. [1]

_____

_____

Base your answer to question 60 on the information below and on your knowledge of biology.

Bacterial resistance to antibiotic treatment is becoming an increasing problem for the medical community. It is estimated that 70% of bacteria that cause infections in hospitals are resistant to at least one of the drugs used for treatment. Dangerous strains of tuberculosis (TB) have emerged that are resistant to several major antibiotic drugs. While drug-resistant TB is generally treatable, it requires much longer treatments with several antibiotics that are very expensive.

60. Explain the loss of effectiveness of antibiotic drugs. In your explanation, be sure to:
  • identify the genetic event that resulted in the original antibiotic resistance in some strains of bacteria [1]
  • explain how the overuse of antibiotics can increase bacterial resistance [1]

_____

_____

_____

_____

_____

Base your answers to questions 61 and 62 on the information below and on your knowledge of biology.

The average life expectancy of humans in the United States increased from 63.3 years in 1943 to 77.6 years in 2003. This, combined with other factors, has led to an increase in population.

61. State *one* factor that contributed to the increase in life expectancy in the United States. [1]

_____

_____

62. State *one* way the increase in population affects other species. [1]

_____

_____

63. The diagram below represents a cell found in some complex organisms. The enlarged section represents an organelle, labeled X, found in this cell.

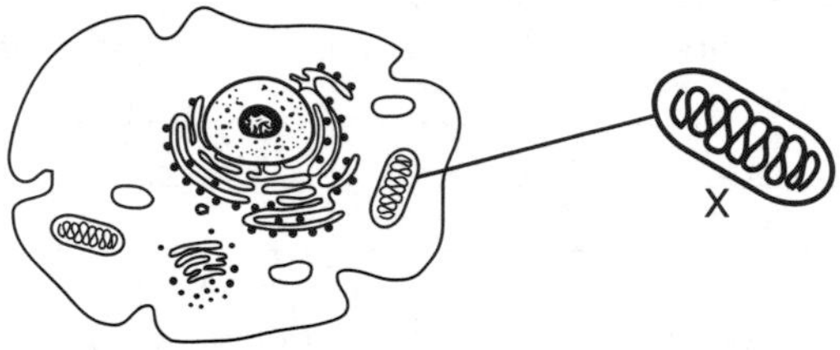

Describe the function of organelle X and explain how it is important to the survival of the cell. In your answer, be sure to:
- identify organelle X [1]
- state the process that this organelle performs [1]
- identify the *two* raw materials that are needed for this process to occur [1]
- identify *one* molecule produced by this organelle and explain why it is important to the organism [2]

_____

_____

_____

_____

_____

_____

_____

_____

_____

_____

_____

_____

Base your answer to question 64 on the passage below and on your knowledge of biology.

The Arctic National Wildlife Refuge

The Arctic National Wildlife Refuge (ANWR) in Alaska is the last great wilderness in America. Many migratory animals stop there to feed and rest. This region also supports an abundance of wildlife, including various types of vegetation, herbivores such as musk oxen and reindeer, and carnivores such as polar bears and wolves.

64. Wolves often hunt reindeer for food. State the effect on the size of the wolf population if the amount of vegetation were to drop suddenly. Support your answer. [1]

_____

_____

_____

Part D

Answer all questions in this part. [13]

Directions (65–77): **For those questions that are followed by four choices, record in the space provided the *number* of the choice that, of those given, best completes the statement or answers the question. For all other questions in this part, follow the directions given in the question and record your answers in the spaces provided.**

65. The amino acid sequences of three species shown below were determined in an investigation of evolutionary relationships.

Species A:	Val	His	Leu	Ser	Pro	Val	Glu
Species B:	Val	His	Leu	Cys	Pro	Val	Glu
Species C:	Val	His	Thr	Ser	Pro	Glu	Glu

Based on these data, which *two* species are most closely related? Support your answer. [1]

_____

_____

_____

June 2010

66. A student carried out a lab activity where she was asked to squeeze a clothespin as many times as she could in one minute and record that number. She immediately tried the same activity again, thinking she could do better the second time, but the number was lower. She immediately tried again, but the number was lower still.

State *one* reason why she continued to get lower numbers, even though she tried to increase the number of squeezes several times. [1]

_____

_____

_____

67. A laboratory setup using an artificial cell made from dialysis tubing is shown in the accompanying diagram. Identify the process that would most likely be responsible for the movement of glucose from inside the artificial cell to the solution outside of the cell. [1]

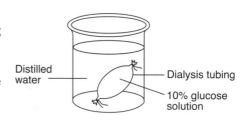

Distilled water

Dialysis tubing

10% glucose solution

_____

Base your answers to questions 68 through 71 on the information below and on your knowledge of biology.

Scientists attempted to determine the evolutionary relationships between three different plant species, *A*, *B*, and *C*. In order to do this, they examined the stems and DNA of these species. Diagram 1 represents a microscopic view of the cross sections of the stems of these three species. DNA was extracted from all three species and analyzed using gel electrophoresis. The results are shown in diagram 2. Based on the data they collected, they drew diagram 3 to represent the possible evolutionary relationships.

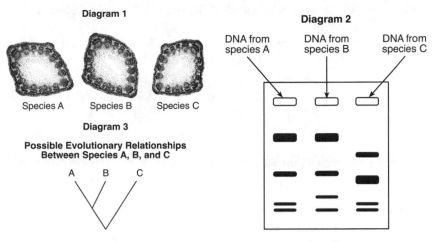

Diagram 1

Species A   Species B   Species C

Diagram 3

**Possible Evolutionary Relationships
Between Species A, B, and C**

A   B   C

Diagram 2

DNA from species A   DNA from species B   DNA from species C

68. State why the evolutionary relationships shown in diagram 3 are *not* supported by the data provided by the stem cross sections in diagram 1. [1]

_____

_____

_____

69. Explain how the DNA banding pattern in diagram 2 supports the evolutionary relationships between the species shown in diagram 3. [1]

_____

_____

_____

70. This technique used to analyze DNA involves the
(1) synthesis of new DNA strands from subunits
(2) separation of DNA fragments on the basis of size
(3) production of genetically engineered DNA molecules
(4) removal of defective genes from DNA   70 _____

71. Explain why information obtained through DNA analysis is a more reliable indicator of evolutionary relationships than observations of stem cross sections with a microscope. [1]

_____

_____

_____

Base your answers to questions 72 through 74 on the information below and on your knowledge of biology.

A wet-mount slide of red onion cells is studied using a compound light microscope. A drawing of one of the cells as seen under high power is shown below.

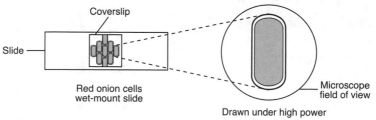

Coverslip

Slide

Red onion cells
wet-mount slide

Microscope
field of view

Drawn under high power

72. On the accompanying diagram, label the location of each of the cell structures listed. [1]

cell wall
cytoplasm
cell membrane

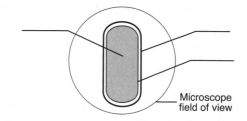

Microscope
field of view

73. Describe the proper way to add a saltwater solution to the cells without removing the coverslip. [1]

_____

_____

_____

74. In the space below, sketch how the cell would look after the saltwater solution is added to it. [1]

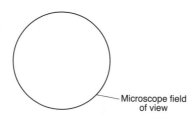

Microscope field
of view

75. A student added an enzyme to a test tube containing a sample of DNA. After a period of time, analysis of the DNA sample indicated it was now broken into three segments. The purpose of the enzyme was most likely to
(1) cut the DNA at a specific location
(2) move the DNA to a different organism
(3) copy the DNA for protein synthesis
(4) alter the DNA sequence in the segment   75 _____

Base your answers to questions 76 and 77 on the diagram below and on your knowledge of biology. The diagram shows the heads of four different species of Galapagos Islands finches.

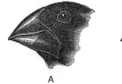  

A   B   C   D

Source: http://Darwin-online.org

76. The four different types of beaks shown are most likely the result of
(1) gene manipulation
(2) natural selection
(3) unchanging environmental conditions
(4) patterns of behavior learned from parents   76 _____

77. Scientists observed that when two closely related species of predatory birds live in different areas, they seek prey early in the morning. However, when their territories overlap, one species hunts at night and the other hunts in the morning. When these two species live in the same area, they apparently modify their
(1) habitat   (3) ecosystem
(2) niche   (4) biodiversity   77 _____

LIVING ENVIRONMENT

ANSWERS
AND
EXPLANATIONS

June 2007
Part A

1. 3   Fungi act as decomposers. Decomposers break down chemicals and nutrients from dead and decaying organisms, recycling them back into the ecosystem.

2. 2   The levels of organization start with the basic unit of life, the cell and build from that. Tissues are composed of groups of cells, while organs are composed of groups of tissues. Organ systems, such as the digestion system, are made of various organs. The diagram in answer 2 shows this relationship.

3. 4   A rapid rise in the number of red blood cells would create a disruption in the circulatory system. Homeostasis is the maintenance of balance within a living organism. The increase in red blood cells would disrupt that balance and thus homeostasis.

4. 1   Proteins found on the surface of cells and viruses that allow for attachment based on *shape* are known as receptors. Receptor molecules play an important role in the immune system. See question and answer # 26 which provide another example of a receptor.

5. 4   Both contractile vacuoles and kidneys have a similar function: maintaining water balance. Although found in different organisms, both structures function to maintain homeostasis (water balance) within that organism.

6. 2   Although genetic material in clones may be identical, the expression of that material can be influenced by the environment. Genes can be "turned on or off" by various environmental factors such as light or temperature.

7. 1   A change in base unit sequence (like TAA → TAC) will lead to a change in the DNA and may alter proteins. These alterations or mutations can lead to genetic variation.

8. 3   In asexual reproduction, the resulting offspring are identical to the parent organism. Budding results in two new yeast cells where the cytoplasm has divided unequally but the genetic material (DNA) is identical.

9. 4   Proteins are composed of building blocks called amino acids. The amino acids link together in a particular sequence to give the protein a specific shape. In proteins, shape relates to function. So, if proteins are shaped or folded differently, they will have different functions.

10. 4   All body cells contain the same DNA, however, different genes are expressed or "turned on" in different types of cells. These genes code for different proteins which give the cells different functions.

11. 2   Genetic engineering is a process where genetic information is removed or "cut" from the DNA of one organism and inserted or "spliced" into the DNA of another. The organism with the inserted gene now has the capability to perform the function encoded in the inserted gene, in this case, frost protection.

12. 3   Mutation in the ovary (3) and testes (6) would provide the most impact on human evolution. These mutations would affect sex cells and may be passed to the next generation. This mutation or variation could lead to evolutionary processes in following generations.

13. 2   During sexual reproduction, there is a recombination of genetic material which leads to different combinations of genes. Also, during the formation of sex cells (meiosis), genes are "shuffled" into sex cells leading to different combinations in each sex cell. These different combinations will lead to variation in offspring.

14. 3   A species with little genetic variation will not have much ability to adapt to change in the environment. If a species is unable to adapt, that species could become extinct.

15. 1   Both the skin cell and fertilized egg cell have the full number of chromosomes (2n). All other answers contain at least one sex cell which only contains half the number of chromosomes (n) compared to a body cell.

16. 4   After a zygote is formed, rapid cell division takes place. This cell division is the same as the process of mitosis. At a certain point in the development of the zygote to an embryo, certain cells begin to become programmed to perform certain functions. This process is known as differentiation.

17. 4   Within the uterus of a human female, a placenta will develop to serve as the exchange site between mother and developing fetus. One substance that passes from mother to fetus is oxygen. All other answers do not describe the human female reproductive system.

18. 1   Structure A is the uterus. The uterus lining is regulated and maintained by two female hormones: estrogen and progesterone. By maintaining the uterus lining, the human female provides a suitable environment for fetal development. In this diagram, D is the fallopian tube, B is the vagina and C is the urinary bladder.

19. 3   Energy in molecules can be found within the chemical bonds of these molecules. Remember that the high energy molecule ATP stores energy in its bonds between phosphate molecules. When these bonds are broken, the energy for necessary life processes is released.

20. 1   The process of cellular respiration produces ATP molecules within the mitochondria of the cell. This energy is stored in the chemical bonds of ATP molecules. Remember that cells use simple sugars like glucose and oxygen to produce ATP molecules.

21. 3   Vaccines contain weakened or heat-killed microbes that when injected into the human body initiate an immune response. An immune response will activate white blood cells to attack infected cells using specific molecules and mark microbes for destruction with other molecules (antibodies).

22. 3   Natural selection is the adaptation of a species to a certain environment based on a selection process that allows the best fit to survive. In this case, the type of feet for each bird species has allowed that bird to survive in their environment.

23. 3   Diagram 3 best illustrates the process of fertilization in which a cell is formed that has all the genetic information needed for the organism to perform life processes. The resultant cell in diagram 3 has the same number and kinds of chromosomes as the original nucleus, $<<$ ll, and each sex cell has one half the original number, $<$ l. When the sex cells combine, they restore the chromosome number of the original cell. $< 1 + < 1 = <<$ ll.

24. 2   In plants, the process of photosynthesis uses the energy of the Sun to convert carbon dioxide into the chemical energy of sugar and give off oxygen as a by-product.

25. 4   Glucose, an energy molecule, is produced by photosynthesizers. Producers directly gather energy from the Sun and transfer it through an energy pyramid to consumers within an ecosystem.

26. 2   Molecule A represents a receptor molecule on the surface of a cell. Molecule B represents a hormone. Hormones interact with receptors in a lock and key mechanism based on particular shapes. When the hormone interacts with the receptor molecule, it activates a cellular response.

27. 4   Energy that is not transferred through an energy pyramid is lost as heat (X) to the environment. As you move from the producers nutrition level to herbivores and carnivores, the amount of available energy decreases.

28. 1   This ecosystem may be unstable because the number of producers (algae) is far less than the number of primary consumers (plankton). With less algae or producers present, less energy is available to be transferred up the energy pyramid in this eco-system.

29. 2   Using solar energy instead of fossil fuels to heat household water would reduce the consumption of non-renewable resources. Solar energy could be used and provides an alternative to using non-renewable resources as an energy source. Remember that non-renewable resources such as coal or petroleum products can't be replenished.

30. 4   The rabbits brought to Australia were an invasive species. They had no natural predators, found the environment suitable and produced large numbers of offspring. With little to no limits on the species, the population grew at a rapid rate.

Part B-1

31. 2   The earthworm measures 90 mm. Using the metric ruler, you can see that the earthworm extends from 1.6 cm to 10.6 cm or a length of 9 cm. Remember that 1 cm = 10 mm, so 9 cm = 90 mm.

32. 3   According to the chart, the smaller the crocodile, such as Group *A*, the smaller their prey. In the column for group *A*, spiders, frogs and insects are the smallest of the prey types. As the crocodile size increases from Groups *A* to *C*, so does the size of the prey, moving to reptiles, fish and mammals.

33. 3   The number of A bases will be 3. On the top section of DNA there is one A. Both boxes that are opposite of a T will also contain A's. Remember that in DNA, the base T pairs with the base A.

34. 4   In section II, the rate of reproduction is greater than the death rate for the bacteria population. The slope of section II represents an increasing population (number of individuals). Based on this increase, there are more individuals being born than are dying.

35. 4   The rectangular shape would be most closely related to the spiral shape. They both belong to the same genus, Felis. As one moves down the classification system from Kingdom (animal) to Genus (Felis), organisms are more closely related.

36. 1   Species A and B will not compete for nesting sites. Species *A* builds nests 1 m – 5 m above ground. Species *B* builds its nests over 10 m above ground. These species will not occupy the same nesting areas. All other answers require information not found in the data table.

37. 1   Survival of the fittest would result from the spraying of the pesticide on an insect population for 3 generations. These darker insects were better adapted to survive the pesticides and went on to reproduce, slowly increasing the dark population number.

38. 3   Diagram 3 shows the relationship between the organisms. Tuna feeds on herring, which feeds on sand eels and seabirds feed on cod which feed on sand eels. Fisherman disrupted this food chain by reducing the number of herring and sand eels, thus decreasing the food source for tuna and cod and indirectly seabirds

39. 1 Sperm are produced in *A* (testes), travel through *C* (vas deferens) and leave the male body through *G* (urethra).

40. 2 Structures *B* and *E* represent seminal glands that secrete fluids that aid in sperm transport. The fluid provides a means of transport and nutrition.

41. 2 Structure *G* is the urethra. The urethra functions to send sperm out of the male body (reproduction) and also functions to direct urine out of the body (excretion).

42. 4 Neither food chain *A* nor *B* use decomposers to supply energy. Decomposers are recyclers of materials and nutrients, they do not supply energy to food chains.

Part B - 2

43. Answer: Chloroplasts

Explanation: Chloroplasts, shown as the small darkened ovals, are the cell organelle that carry out the process of photosynthesis or autotrophic nutrition.

44. Answer: Ribosome

Explanation: Ribosomes (the small dots) can be located on the channels of the endoplasmic reticulum that surround the nucleus. Remember that ribosomes are the site of protein synthesis.

45. 3 All cells use energy that is released from ATP. ATP is produced in cell organelles called mitochondria through a process of cellular respiration.

46. 1 The diagram shows the energy relationship between organisms. As one moves from the producers such as trees and grass, energy flows to the consumers like mice and crickets. The arrows represent the flow of that energy.

47. 3 Grass is the producer that carries out photosynthesis or autotrophic nutrition. Remember that photosynthesis converts carbon dioxide and water into sugar in plants using the Sun's energy.

48. Answer: The cricket population would decrease.

Explanation: Crickets feed on grass. If grass was removed, crickets would lose their food source and their population numbers would decrease.

49.

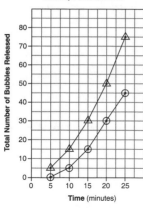

The Effect of Temperature on Respiration in Yeast

Key
⊙ Yeast respiration at 20°C
△ Yeast respiration at 35°C

Total Number of Bubbles Released (y-axis: 0, 10, 20, 30, 40, 50, 60, 70, 80)

Time (minutes) (x-axis: 0, 5, 10, 15, 20, 25)

Explanation: The scale for the Time axis should be set up in intervals of 5 with every two spaces representing 5 minutes. The Total Number of Bubbles Released scale would be set up in intervals of 10, with two spaces representing 10 bubbles

50-51. See above Graph

Credit will only be given if circles or triangles surround the appropriate point and no credit is given if points are plotted that are not in the data chart.

52. Acceptable answers include but are not limited to:
— As temperature increases, the gas production increases.
— As temperature decreases, the gas production decreases.
— There is a direct relationship between temperature and gas production.

Explanation: Based on the data table and as shown in the graph, at every time interval (minutes), the number of bubbles released was more for the 35°C water bath setup than for the 20°C water bath setup.

53. Answer: Carbon dioxide (CO_2)

Explanation: Cellular respiration in yeast produced the gas, carbon dioxide, as a result of the oxidation or breakdown of glucose.

54. Answer: Letter X is the pancreas.

Explanation: The pancreas produces the hormone, insulin, which regulates blood sugar. In the diagram, organ X, the pancreas, produces insulin.

55. 2   A feedback mechanism regulates the production of a hormone. The dashed line represents feedback information to the pancreas (X) that would reduce or increase the production of insulin as needed. This process maintains homeostasis.

Part C

56. Hypothesis: *Acceptable responses include but are not limited to:*
    — Competition decreases plant height
    — Competition increases plant height
    — Competition has no effect on plant height

    Same factor: *Acceptable responses include but are not limited to:*
    — same soil type or amount
    — same environmental conditions such as sunlight, water, temperature
    — same type of plants

    Dependent variable: height or size of plant

    Data: *Acceptable responses include but are not limited to:*
    — the data supports my hypothesis because plants in the pot
       with the greatest number of plants are the shortest.
    — the data does not support hypothesis because plants in
       Pot *C* (20 plants) are shorter than plants in Pot *A* (5 plants).
    — Data does not support my hypothesis because the number
       of plants did affect the height or size of the plants.

    Explanation: Most experiments are carried out to test hypotheses. A hypothesis
    is a suggested explanation to a problem. In this case, how competition affects
    plant height. In a valid experiment, there should be only one variable,
    all other factors must be kept the same to provide valid data. The
    dependent variable is the factor that is changed based on the manipulation
    of another factor (variable). In this experiment, the height or size of the
    plant depends on the number of plants in each pot (independent variable).
    The data collected clearly shows that competition does affect plant height
    or size. The pot with the most plants (Pot *C*) had the shortest plants and
    Pot *A* with the fewest plants and less competition had the tallest plants.

57. Life processes affected by pH:
    Acceptable responses include but are not limited to:
    — Growth   — Digestion   — Reproduction   — Transport

    Environmental problem: *Acceptable responses include but are not limited to:*
    — Acid rain   — Loss of biodiversity   — Habitat loss

    Cause of Problem: *Acceptable responses include but are not limited to*:
    — Air pollution   — Burning of fossil fuels   — Deforestation

    Explanation: pH change could affect many life processes by altering the shape
    of proteins. These proteins may be responsible for performing many vital
    functions. One type of protein that could be affected would be enzymes.
    An enzyme speeds up a chemical reaction such as the breakdown or
    synthesis of a molecule. Because the digestive system relies heavily on
    the chemical action of enzymes, digestion as a life process could be very
    much impacted by a pH change. Ecosystems that are stable, operate
    within a narrow range of pH values. If that pH is altered, the ecosystem
    is disrupted leading to loss of habitat or organisms. An example of
    an environmental problem related to pH is acid rain. Acid rain is caused
    by the release of sulfur, nitrogen and carbon dioxide compounds into
    the atmosphere as pollutants. These compounds change the pH of the rain
    that falls in certain areas, usually northeast of where they are released.

58. Introduction: *Acceptable responses include but are not limited to:*
    — Zebra mussels and gobies were introduced into the Great Lakes from the ballast tanks of cargo ships.

    Problem: *Acceptable responses include but are not limited to:*
    — Zebra mussels clog water intake pipes.
    — Zebra mussels disrupt existing food chains.
    — Gobies eat the eggs and young of other fish.

    Increased PCB's: *Acceptable responses include but are not limited to:*
    — Zebra mussels filter PCB's from lake water. Gobies eat zebra mussels then are eaten by sport fish.

    Explanation: Zebra mussels and gobies are invasive species which were introduced to the Great Lakes from the Caspian Sea via the ballast water of cargo ships. These invaders have no natural predators and their populations grow unchecked. The zebra mussels disrupt natural food chains and also clog water pipes of power plants and factories creating a financial burden. Gobies also disrupt food chains by feeding on other fish and their eggs. These invasive species also have contributed to an increase in the concentration of PCB's, harmful chemicals, in sport fish. Zebra mussels are filter feeders and filter PCB's from the lake water. The gobies feed on zebra mussels and take in the PCB's from them. When sport fish feed on the gobies, they in turn ingest more PCB's from the gobies. This is known as bio-magnification.

59. Improved medicine: *Acceptable responses include but are not limited to:*
    — genetic test to diagnose disease
    — gene therapy
    — genetic engineering to produce hormones
    — understanding causes of inherited diseases
    — prevent diseases

    Problem: *Acceptable responses include but are not limited to:*
    — screening for genetic diseases may limit insurance coverage
    — gene therapy could result in overpopulation
    — may lead to discrimination

    Explanation: Research on genetic material has led to our understanding of how genes function and techniques or procedures that can alter genes or prevent genetic expression. Genetic engineering and gene therapy are two techniques that have been used to correct diseases by either altering the genes within an organism or creating a missing protein not originally coded for. Once an understanding of the actual function of genes was determined, that allowed geneticists to design tests to diagnose genetic disorders and even develop tests to prevent certain genetic diseases. With all new knowledge, there will always be concerns that must be addressed. Personal genetic knowledge could encourage insurance companies to deny coverage or assign large premiums. This knowledge could also lead to discrimination if it were made public or available to employers.

60. Environmental concerns: *Acceptable responses include but are not limited to:*
   — chemical may not be biodegradable
   — chemical may interfere with the food webs
   — chemical may pollute environment
   — product may be toxic to humans or wildlife

   Explanation: When a new product is introduced into the environment, all possible effects to the environment and the organisms in that environment must be considered. It must be determined how the chemical will affect other living organisms (toxicity) in that area as well as the affect on the food chain. The chemical will need to be tested to see if it breaks down in the ecosystem and if it will become a pollutant. Many tests are required by companies during research and development of new chemicals to make sure that chemicals are safe for the environment.

61. *Acceptable responses include but are not limited to:*
   — wear goggles *or* shoes *or* gloves *or* masks
   — follow directions on package
   — do not spray on non-target areas

   Explanation: To prevent exposure to the chemical, the sprayer should wear appropriate safety gear to protect skin, eyes and airways. The directions on the package provide instructions for safe application of the chemical and should be followed. These directions have been developed to provide the safest means of applying the chemical with the least amount of danger to the sprayer or the environment.

Part D

62. 3   An accurate experiment should have only one testable variable. The addition of another clothespin which is difficult to open creates two variables: exercise vs. resting and easy/difficult clothespins. One could not determine if it is the exercise or the clothespin type that affects the number of squeezes. All other answers are parts of a valid experimental set up.

63. 2   This technique, used to analyze DNA, is called gel electrophoresis. Gel electrophoresis separates various size fragments of DNA as they travel through a gel. Smaller pieces travel farther though the gel than larger pieces, thus displaying bands of separated DNA, each varying in size.

64. 1   The process of separating fragments of DNA is called gel electrophoresis. An electrical charge is sent through the gel which causes the DNA to migrate through the gel. The DNA moves varying distances based on the size of the fragment. (See answer for question 63).

65. *Acceptable responses include but are not limited to:*
— solving crimes   — determining evolutionary relationships
— paternity testing   — determining identity
— gene testing for diagnosis

Explanation: Based on the knowledge that everyone has their own unique DNA, this technique has been used to identify or eliminate suspects in a crime based on evidence. Comparisons can be made with a suspect's DNA and crime scene evidence. Similarly, DNA gel electrophoresis can eliminate or implicate someone in a paternity case by matching parts of a baby's DNA to that of the father.

66. 4 Controlled experiments have one independent variable. This variable will influence the dependent variable. An example of an independent variable would be time.

67. 1 As heart rate increases, the rate of circulation will also increase leading to an increase of blood flow to the muscles. This increased blood flow will transport vital nutrients to muscles including oxygen, which muscles will use to produce energy.

68. *Acceptable responses include but are not limited to:* Cactus finch because it eats mainly plant food whereas the other two finches eat mostly or all animal food.

Explanation: Competition occurs when two organisms try to occupy the same niche. Remember that a niche is a role that an organism plays in an ecosystem. On this island, there should be little to no competition between the cactus finch and the warbler and woodpecker finch. Using the Variation of Beaks Wheel, you can see that they feed on completely different food sources. Since the cactus finch has a different food source that means it occupies a different niche and therefore there would be no competition with the other finches.

69. 1 Due to a decrease in the numbers of small seeds during the dry years, those finches with the larger beak trait were able to survive and reproduce. The large beak trait was then passed on to future generations. This is an example of natural selection.

70. 4 Due to a diminished amount of seeds, there will be competition for whatever food is available. Those finches that have a better adapted trait, such as a larger beak, can out-compete other finches and survive, and then reproduce and pass that trait on.

71.
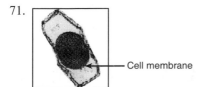
Cell membrane

Explanation: When salt water was introduced to the slide, the amount of water in the onion cell was greater than that of the salt water outside the cell. Diffusion of water or osmosis occurred. Water moved out of the cell and the onion cell became plasmolyzed when the cell membrane pulled away from the cell wall as the cell lost water.

72. 4   Small molecules are able to pass through the dialysis tubing membrane. In order for the green color to be present both inside and outside the tubing, both the yellow and the blue food coloring must have moved. This would suggest that both molecules are small enough to pass through the membrane.

73. 3   The movement of molecules across a membrane from a high concentration to a lower concentration is known as diffusion. Diffusion is a passive transport process requiring no energy. At the beginning of the experiment, there is a higher concentration of blue food coloring inside the tubing and a higher concentration of yellow outside. Both food colorings will move down a concentration gradient from high to low across a membrane. In other words, blue will move out and yellow will move in.

June 2008
Part A

1. 1   Owls, cats and sharks all carry out heterotrophic nutrition, that is, they consume preformed organic substances. Rows *B* and *C* each contain an autotrophic organism (*B*- corn and *C*- alga) which synthesize organic compounds.

2. 4   Stable ecosystems contain more producers which create a solid base for the energy pyramid. An ecosystem with more consumers than producers would experience instability because there would not be enough nutrients to support the number of consumers.

3. 1   While all cells in humans have identical genetic makeup, certain genes in certain cells are "turned on" or expressed during a process called differentiation. Each cell's function is determined by the specific set of genetic instructions that are expressed.

4. 4   Single-celled organisms have various cell parts or organelles that perform essential life processes. These cell parts carry out similar functions as those of human organ systems. For example, a lysosome containing enzymes that digest food particles could be compared to the human digestive system.

5. 3   The mitochondria's primary function is the production of energy, in the form of ATP, from nutrients. A disruption of the mitochondrial process would result in the cell not being able to release this energy.

6. 1   The environment can play a role in gene expression or which genes are "turned on". In the case of bread mold, the environmental factor of temperature plays a role in the expression of genes that control color. In warm temperatures, the bread mold is dark and in cooler temperatures, the mold is red.

7. 1   In asexually reproducing organisms, hereditary information is passed from a parent organism to resulting offspring through DNA. DNA consists of coded instructions using sequences of 4 nitrogen bases – A,T,C, and G. The sequence of these four bases determines the structure of necessary proteins in new organisms.

8. 4   Bacteria reproduce by binary fission, an asexual form of reproduction. This type of reproduction is rapid and results in many offspring in a relatively short period of time. Any adaptive change or mutation therefore will evolve more quickly based on this high rate of reproduction.

9. 2   Amino acid molecules are simple building blocks that are used to synthesize more complex molecules called proteins. These complex molecules of protein then could be used for a structural or digestive function.

10. 2   The gene (*D*) would be the smallest structure located within a chromosome (*A*). Chromosomes are found in the nucleus (*B*) of a cell. The nucleus is an organelle or a smaller part of the cell (*C*). Diagram 2 reflects this relationship.

11. 3   The process represented is that of DNA replication. DNA, being a double helix, must untwist to separate the two strands so that new portions of DNA bases can be constructed. The result of the replication process is two identical DNA molecules.

12. 4   The process of choosing animals with desirable characteristics or traits for breeding purposes is called selective breeding. This process is used to produce not only dog breeds but many desirable agricultural animals and plants.

13. 2   Natural Selection is a process where the best-adapted organism for the environment survives and passes on that adapted trait. In this case, the insects resemblance to the bark of the tree was an adaptive advantage for survival. Those that blend with the tree bark had a higher survival rate and passed on that trait to their offspring.

14. 4   When the environment changes, an organism that lacks adaptive traits for survival and reproduction will die out, becoming extinct. If members of the entire species lack an adaptive trait, the species itself will die out becoming extinct.

15. 3   Photosynthesis and respiration both involve the following molecules: glucose which is organic, and water, oxygen, and carbon dioxide which are inorganic. Photosynthesis uses CO_2 and water to produce glucose and O_2. Respiration uses glucose and O_2 to produce CO_2 and energy.

16. 3   By recombining genes through reproduction, new combinations of genetic information can occur. These new gene combinations may result in genetic variation and may be inherited through gametes during sexual reproduction.

17. 2   Cell 1, a sperm, and cell 2, an egg, were both formed by the process of meiosis. Meiosis is the formation of sex cells or gametes. In humans, cell 1 is formed in the testes, while cell 2 is formed within the ovary.

18. 1   The placenta is the site for exchange for nutrients and wastes between the mother and developing embryo. Mammals without a placenta must use alternative means to supply the nutrients to their embryo. Kangaroos have a pouch where nutrients are provided for the developing embryo.

19. 1   ATP is a high energy molecule with energy being stored within its phosphate bonds. Animal cells use ATP for many uses including the synthesis of materials.

20. 2   Persons receiving an organ transplant should take medication that reduces their immune response to prevent their body from attacking and rejecting the foreign proteins in that organ. Remember, that your immune system attacks any foreign protein or antigen that enters the body.

21. 1   By definition, a clone is an exact genetic copy, so the cells of the new plant should each have the same number of chromosomes and the same types of genes.

22. 2   The developmental changes in the embryo are a result of differentiation and growth. Individual cells are genetically programmed to develop into specific organs or tissues. Cell division will increase the size of the embryo.

23. 3   White blood cells are large cells with distinct nuclei that are capable of engulfing pathogens. In the diagram, Cell A represents a white blood cell that is engulfing smaller particles which may represent disease-causing organisms. This engulfing process is known as phagocytosis.

24. 1   Energy availability in an ecosystem is dependent on producers. Producers, or autotrophs use light energy and convert it into the chemical energy of organic molecules. This process is known as photosynthesis. Organisms that consume these organic molecules then gain this energy for their use in growth and survival.

25. 2   Ecosystems are based on the interaction of plants and animals with their physical environment. Climate conditions is the factor that most influences which type of plants or animals can survive in any given geographic area. If a climate is hot and dry, a desert ecosystem will form with specific types of plants and animals.

26. 2   A trade off occurs when something is given up in order to gain something else. In the case of farming, by planting a single crop such as corn or wheat, humans lose biodiversity but gain valuable crops for feeding the world's populations.

27. 4   Grass is the single producer in the food chain. Crickets, which feed on grass, act as herbivores and the frog and owl are both predators hunting for prey. The owl preys upon frogs and the frog preys upon crickets.

28. 3   Ecological succession is a natural process where an ecosystem undergoes change in plant life based on the amount of soil that has formed. As more soil is formed, the size of the plants increase, from smaller plants to large trees. The type of animal life will also change in response to the changing plant life.

29. 1   Through the use of technology, humans have the ability to modify their environment. Humans can change landscapes, remove forests, and create destructive environmental pollutants. These changes have a great impact on the other species that inhabit these areas. These species are unable to modify their environment and must either adapt, move or face extinction.

30. 3 The growth of the rabbit population is a direct result of successful competition. Rabbits reproduce quickly, can live in many habitats and are fast runners, avoiding predation. All of these factors have contributed to their successful competition for food with the native species.

Part B-1

31. 3 The lab procedure represented in the diagram involves the introduction of stain to a wet mount slide. The stain is carefully added by the dropper to one side of the cover slip while a paper towel on the other side absorbs water. This absorption process pulls the stain across the slide under the cover slip without disturbing the cover slip.

32. 2 The insertion of a foreign gene onto human DNA raises many ethical concerns. Scientists and researchers must answer many questions, discuss advantages/disadvantages and play out many scenarios before this experimentation should be done.

33. 4 A conclusion cannot be reached until an experiment has been carried out. It would be based on collected data and results of a well-planned experiment. The experiment should be duplicated more than once to validate the results and its supportive conclusion.

34. 3 Hormones are chemical messengers secreted by glands that are transported to specific target organs by the bloodstream. These hormones cause a change to occur that sometimes can affect a gland.

35. 1 The student would have to assume that starch is synthesized from the glucose produced in the green chlorophyll areas. Photosynthesis converts light energy into the chemical bond energy of glucose. Plants then use glucose to synthesize storage molecules of starch. The green chlorophyll areas of the leaf can carry out photosynthesis while the white areas, lacking chlorophyll cannot. Therefore the white areas will not contain starch synthesized from glucose.

36. 1 Structure A in both Cell X and Cell Y is the cell or plasma membrane. The cell membrane serves as a selective exchange site allowing for the movement of nutrients and waste products. Metabolic wastes would diffuse out of the Cells X and Y through structure A.

37. 4 Each species of bacteria - A, B, C, and D has a specific range of temperatures, represented by the arch shape, in which it can successfully reproduce. Some of the ranges overlap but each species has an optimum temperature for ideal growth that differs from that of the others.

38. 2 Genes contain the coded instructions for making strands of mRNA. The codes are sections of DNA each with a specific sequence or order of nitrogen bases (A,C,T,G) that direct the synthesis of a particular protein.

39. 4   The base of the energy pyramid is made up of producers with the highest amounts of energy. As one moves up the energy pyramid, the energy content goes down. The bar graph in answer 4 represents this relationship.

40. 1   Genetic diversity or variation provides for the chance that a trait might help an organism adapt and survive a change in its environment. By adapting, the organism would be able to reproduce and allow for survival of the species.

41. 3   Based on the diagram, earthworms and sea stars have a common ancestor. This common ancestor is found directly up from the ancestral protists where the straight line branches left and right. The left branch leads to the earthworm and the right branch leads to the sea star. By sharing a common ancestor, the earthworm and sea star are evolutionarily related.

42. 2   Day 8 represents the day where the carrying capacity was reached. Carrying capacity is defined as the population number that can be maintained by the available resources in that environment. At day 8, the population is no longer growing or increasing and has leveled off. This indicates that the carrying capacity has been reached and the population has reached a level where the available resources can maintain but not increase the population.

Part B-2

43 and 44

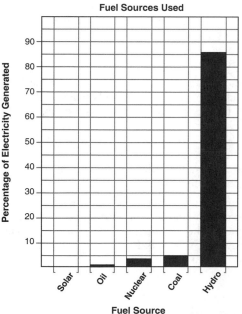

Fuel Sources Used

43. Explanation: Each horizontal line should represent 5 percent. The scale represented starts numerically at 10 on the second line, 20 in the fourth line and so on.

44. Explanation: On the graph, the height of the vertical bars should match the data given in the data chart. The bars should be shaded in.

45. Answer: Coal *or* Oil

Explanation: Both of these fuel sources are a type of fossil fuel that was formed over many years from organic materials buried within the earth's crust.

46. Answer: Hydro *or* Solar

Explanation: Each of these energy sources has an unlimited supply and therefore is renewable. They cannot be used up. Remember that nonrenewable resources have a limited supply and are not replaceable.

47. Acceptable answers include but are not limited to: Burning coal can cause produce air pollution *or* acid rain *or* global warming.

Explanation: The burning of coal releases chemicals and carbon dioxide into the air. The chemicals can lead to the pollution of the atmosphere or the creation of acid rain. The release of carbon dioxide creates a greenhouse effect and leads to global warming.

48. Acceptable answers include but are not limited to: The hawk population will decrease because there will be fewer snakes since there are fewer frogs for them to eat.
or The hawk population will increase because there will be more grasshoppers for the shrews to eat and more shrews for the hawks to eat.

Explanation: In a food web, arrows represent the flow of energy in a feeding pattern. In this food web there are two feeding patterns to the hawk: The first path – grasshopper to frog to snake to hawk. If the frog is removed, the snake population will decrease due to lack of food. This will, in turn, affect the hawk population negatively causing a decrease. The second path – grasshopper to shrew to hawk. If the frog population decreases, the grasshopper population increases creating more food for the shrews. The shrew population increases providing more food for the hawks, thus increasing their population.

49. Acceptable answers: Chloroplast *or* Cell wall

Explanation: Producers in a meadow ecosystem would belong to the Plant Kingdom. Plant cells contain chloroplasts and cell walls which would not be found in carnivore cells. In this ecosystem, the grass and shrubs are the producers.

50. Acceptable answers include but are not limited to:
- mating with another earthworm allows for variety in the species.
- better chance of survival due to variation or genetic recombination.

Explanation: When two individual earthworms mate with each other, they each contribute one half of the genetic makeup for their offspring. By combining their genetic contributions, the offspring will have a mix or new combination of genes, thus providing variation. Remember that variation is a key component to the process of Natural Selection.

51. Acceptable answers:
5a has light *or* white *or* clear wings
5b has dark *or* black *or* shaded wings

Explanation: In a dichotomous key used to identify organisms, each section or step will have two choices that describe a difference between organisms. In the case of 5a and 5b of Species E and F respectively, the only difference visible is that of their wings. The missing information is given in the answer section above

52.

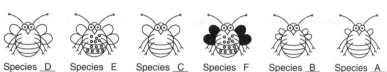

Species _D_   Species E   Species _C_   Species F   Species _B_   Species _A_

Explanation: For this question, you must use the dichotomous key to identify all the unknown insects. Always start each unknown identification at step 1. For example, for the first bug on the left, starting at step 1, we have two choices: small wings or large wings. This bug has large wings so follow the dots to the right where it directs you to go to 3. At step 3, our choice involves number of wings (single or double). This insect has double wings so following the dots to the right, it directs us to go to 4. This insect at step 4 does not have spots so following dots to the right we see that it is species D. Repeat procedure for all unknowns, starting at step 1 for all.

53.

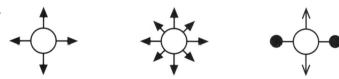

Explanation: The proteins (Y–shaped structures) on the surface of the human cell represent receptors. In order for a virus to attack a cell, it must have proteins whose shape matches that of the receptors on the cell it is attacking. In the case of the diagram, the virus would have to have proteins on its surface in the shape of a triangle (arrow) to match that of the protein receptors (Y) on the human cell.

54. Acceptable answers include:
Mutation *or* Mutagenic agent that led to new protein.

Explanation: A change in protein structure or shape would be caused by a change in the DNA. Any change in DNA is a direct result of a mutation or caused by a mutagenic agent.

55. Acceptable answers include: Parasite /Host *or* Pathogen /Host

Explanation: The virus cannot survive on its own and must use other cells (hosts) to reproduce, damaging the cell as a result. This type of relationship would be termed a parasitic relationship with the virus being the parasite (an organism that benefits) and the cell being a host. The virus is also called a pathogen because it's disease causing.

June 2008
Answer Key

Part C

56. Answer: Pancreas *or* islets of Langerhans

    Explanation: The hormone insulin is produced by the pancreas. Insulin helps to maintain blood sugar levels and thus homeostasis within the human body.

57. Acceptable answers include: Glucose *or* sugar *or* ketones

    Explanation: Insulin's function is to remove sugar or glucose from the blood by making it available for use by cells. When insulin is not secreted in normal amounts, it will affect the concentration of glucose, sugar or ketones in the blood.

58. Acceptable answers include but are not limited to:
    - They can transfer pathogens to humans or domestic animals.
    - Imported species may displace native species.
    - increased competition for food and/or habitat for native species

    Explanation: The importation of hedgehogs as an exotic pet could have several negative effects. Pathogens carried by the hedgehogs such as foot and mouth virus, salmonella and fungi, may be transferred to humans or other domestic animals. If hedgehogs are released into the wild, they may compete with native species of food or habitat. Because of their spiny fur, they may have no natural predators and could even displace native species.

59. Acceptable answers include but are not limited to:
    - Produce antibodies
    - White blood cells will engulf and destroy pathogens.

    Explanation: A pathogen that enters the human body will produce an immune response. Several events will occur during this response. The human immune system will produce T cells that directly attack the pathogen, and produce B cells that are responsible for producing large amounts of antibodies which mark the pathogen for destruction. Finally, the immune system will signal white blood cells to engulf and destroy the pathogens or pathogen infected cells. (Also refer to question and answer for no. 23)

60. Acceptable answers include but are not limited to:
    - The coyote population will decrease because the wolf will be a competitor for the same prey as the coyote.
    -The coyote population will be unaffected because there is sufficient prey for both wolf and coyote.

    Explanation: By introducing the wolf back into the Adirondack Mountains, there may be competition for prey. Since the wolf and coyote have the same prey, the wolf if successful, could out compete the coyote for limited food sources thus reducing the coyote population. However, if there is ample prey, both species could co-exist without any negative effects on the coyote population.

61. Acceptable answer include but not limited to:
-Coyotes control the growth of certain prey populations.

Explanation: A limiting factor is one that influences the growth of a population. It could be biotic (living) or abiotic (non-living). In this case, the coyote is the limiting factor for the prey population because its predatory activity limits the growth of the prey populations such as deer, beaver and moose.

62. Acceptable answers include but are not limited to:
- The wolf was once a natural part of the ecosystem.
- To control deer population
- There is adequate prey to support wolf population
- Increase biodiversity

Explanation: Reintroduction of wolves into the Adirondacks may restore a native species back into the ecosystem. This would increase biodiversity and thus make the ecosystem more stable. There is adequate prey such as deer, beaver, and moose, for the wolf to feed on. In fact, the deer population which is very high might be controlled by the introduction of a new predator.

63. *Hypothesis*
Acceptable answers include but are not limited to:
- Tomato plants exposed to 16 hours of light will grow faster than those exposed to 8 hours of light.
- A brighter light will cause the tomato plant to grow larger.
- Light affects plant growth.

Explanation: A hypothesis is a statement that suggests an explanation for an event or phenomenon. In this case, the effect of light on tomato plant growth. To generate possible answers, the hypothesis must be in the form of a statement and must include light and a form of the growth (height, size, etc.) in the statement.

Independent variable
Acceptable answers include but are not limited to:
-Amount of light
-Light
-Intensity of light

Explanation: The independent variable is the variable that can be manipulated so that it will affect the dependent variable. In this case, light can be manipulated by amount or intensity so that its effects can be measured by plant growth.

Data
Acceptable answers include but are not limited to:
- Height of plants
- Number of leaves
- Mass of plants
- Size of leaves
- Amount of growth

Explanation: Data in this experiment must measure plant growth. The data could include any measurements that are numerical such as centimeters of height or grams of mass.

64. *Function*
Acceptable answers include but are not limited to:
- Gas exchange
- Respiration
- Photosynthesis

Explanation: The guard cells on a leaf surround a stomate and serve to open and close that stomate. The stomata allow for the exchange of gases (CO_2 and O_2) into and out of the plant. This movement of gases allows for important life processes, such as respiration and photosynthesis, to occur by providing necessary ingredients. Remember that water also moves through the stomata during a process called transpiration which drives water movement up the plant.

Carry out function
Acceptable answers include but are not limited to:
- guard cells change shape
- guard cells change size of the leaf opening (close up stomate)

Explanation: Guard cells can change their shape through the movement of water. When a guard cell takes in water, the cell becomes plump and bends like a crescent. This shape change pulls open the stomata or leaf opening. And likewise, when the guard cell loses water, the stomata opening decreases or closes.

Evolution
Acceptable answers include but are not limited to:
- prevents evaporation on a sunny day
- prevents the entrance of some pollutants

Explanation: Guard cells, on the bottom of the leaf, will experience less direct sunlight and therefore lower temperatures. This may be an advantage to land plants so they don't evaporate excess water on hot, sunny days. The location of the guard cells on the bottom of the leaf also may prevent airborne pollutant particles from getting into the stomata from above.

65. Acceptable answers include but are not limited to:
- No, mutations to body cells are not transmitted to offspring.
- No, only mutations to gametes are transmitted to offspring.

Explanation: Any mutation to skin cells due to ozone depletion resulting in more UV exposure will not be passed on to offspring. Gametes are the only cells responsible for the passing of genetic information. The mutation would have to occur to these sex cells in order to get passed to offspring.

66. Acceptable answers include but are not limited to:
    - decrease in consumers or biodiversity
    - decrease in oxygen
    - decrease in available energy
    - increase in carbon dioxide

    Explanation: Photosynthetic organisms function as producers in the ocean ecosystem. Producers convert light energy to chemical energy and pass that along to consumers. With fewer producers, there will be less energy to pass on and therefore consumer populations will be affected in a negative way. With fewer producers and less consumers, biodiversity would also decrease leading to a less stable ecosystem. Producers utilize carbon dioxide during the process of photosynthesis, while releasing oxygen. With a decrease in producers, photosynthesis would decrease resulting in less oxygen and more carbon dioxide in the atmosphere.

67. Acceptable answers include but are not limited to:
    - doesn't allow for recycling of nutrients into the lawn
    - takes up land fill space

    Explanation: When lawn clippings were removed from their original growth site, the nutrients contained in those clippings were removed and not recycled back into the ground. Over time, the lawn could become nutrient deficient. By adding lawn clippings to the landfill, space is taken up by materials that could have been recycled. With an increased need for more landfill space, habitats could be destroyed to fill that need.

Part D

68. 2   Enzymes known as restriction enzymes are used to cut DNA at specific sites to produce fragments of varying lengths. These fragments are then run through gel by electrophoresis for analysis to determine identity, crime scene information or even paternity.

69. 1   Paper chromatography is a process whereby substances within a solvent are separated by traveling up through paper. The molecular weight, solubility and reaction of the substances with the paper will determine how far up the paper they travel. Specific patterns (distances) traveled help to determine identity of these substances.

70. 1   Muscle cramps are a result of the build up of lactic acid within the muscles. This condition occurs when the runner is experiencing an anaerobic condition in the muscles. In an anaerobic condition, there is not an adequate supply of oxygen provided to the muscles.

71. Plant Species – A and C – most characteristics in common
    or   same type of chlorophyll

    Explanation: By examining all of the given characteristics in the chart, the two that show all the same characteristics are species A and C. Because they have the same characteristics, this suggests that they evolved from a common ancestor and are closely related.

June 2008   Answer Key

72. Acceptable answers include but are not limited to:
- protein structure
- types of enzymes present
- DNA sequence
- other physical characteristics

Explanation: Closely related species based on evolutionary relationships would have similar DNA sequences and therefore similar proteins or enzymes. Also they might have similar physical structures such as root or flower structure.

73. Acceptable answers include but are not limited to:
- Two related species may produce similar substances that could be used for medicine.
- A related plant might provide a cheaper source of the substance.
- If a plant becomes extinct, a related plant might provide an alternative source of the substance.

Explanation: Scientists use evolutionary relationships in their research. Many plants provide chemicals for medicinal purposes. By identifying close relationships between plants, researchers might identify more sources for helpful chemicals. With more sources available for helpful chemicals, costs may go down or the second source might be cheaper to obtain. Lastly, if a plant species becomes extinct, closely related species may provide sources for substances lost to extinction.

74. 3 Species B feeds on worms and insects, which are considered to be part of the Animal Kingdom. An organism that consumes other animals is known as a carnivore.

75. 2 Species B and C could live in the same habitat. Species B, as mentioned in answer 74, is a carnivore feeding on worms and insects while Species C is an herbivore, feeding on plant materials such as fruit and seeds. Since they use different food sources, they will not compete for food.

76.  This experiment is designed to show the process of diffusion. Small starch indicator molecules readily diffuse through the dialysis membrane from a high concentration outside to a low concentration inside. However, the starch molecules are too large to diffuse through the dialysis membrane and will not move. After 1 hour, the expected locations of the molecules will show starch indicator molecules both in and out of the dialysis bag and the starch molecules only inside the bag.

77. Acceptable answers include but are not limited to:
- a blue black color indicates the presence of starch
- a color change would occur

Explanation: The starch indicator solution (most likely iodine) would show a color change from amber to blue black when in the presence of starch. Since the starch indicator diffused into the dialysis bag (see question 76), a color change would occur in the bag where it contacted starch molecules.

78. Acceptable answers include but are not limited to:
 - some molecules are too large to pass
 - some molecules are insoluble
 - the permeability of the membrane

Explanation: Molecule movement through a membrane is based on several factors. If molecules are large, they may be too big to move readily through the membrane and may be moved through only by phagocytosis or engulfment. Some molecules move through the membrane because they are lipid soluble. If a molecule is insoluble in lipids, it will not be able to pass. Remember that the plasma membrane is composed of a lipid bilayer. Lastly the membrane itself is selectively permeable to molecules based on size, shape and solubility, it ultimately determines what moves through.

79. 2 To determine the width of the cell, one could determine how many cells would fit across the width of the field of view. Five cells can fit widthwise in the field of view. The total field of view is 4000 μm wide. Divide 4000 μm by 5 cells to get 800 μm, per cell.

80. 3 Diagram 3 shows a plant cell that has lost water and has the inner cell membrane pulled in leaving the cell wall still intact. When a plant cell is placed in a salt solution, water will diffuse out of the cell by osmosis causing the cell membrane to pull in and away from the cell wall. The cell wall still retains its shape and is generally not affected by the salt water, although the cell may be a little smaller.

1. 1 The mature forest community stage of ecological succession generally remains stable over time unless a disturbance changes the environment. This diagram represents succession, the gradual ecological change that occurs within an area, evidenced by changes in soil depth and vegetation type.

2. 4 Oxygen transport is carried out by the circulatory system. Red blood cells which contain hemoglobin transport oxygen throughout the body. The heart pumps the blood containing oxygen to the cellular level.

3. 1 Structure 1 is the nucleus which contains DNA, the genetic material that contains the code for protein synthesis.

4. 1 Specific genes within cells are activated or "turned on" during embryonic development and perform specific genetically coded functions. Remember that all cells within an organism contain the same genetic information.

5. 2 Environmental factors can influence genetic expression. The action of those genes can be modified to create different genetic results. For example, if one twin spent time in the Sun while the other did not, their appearance would differ in skin tone. The Sun exposed twin would have darker skin as those genes would have been expressed through exposure to sunlight.

6. 3 Insulin, produced in the pancreas, is a hormone that regulates blood sugar levels. All other hormones play a role in reproduction.

7. 3 By examining the contents of the owl pellets, students could determine the diet of the owl. Based on the owl's diet, the student could determine where the owl fit into the food web and its role for that ecosystem.

8. 1 DNA base sequences contain codons (3 nitrogen bases) that can code for a specific amino acid. When amino acids are sequenced in a specific order, they determine the shape of a specific protein. A protein's shape then determines its function, which can be expressed as a particular trait.

9. 1 Sexual reproduction leads to the recombination of genetic material and thereby genetic variation. Within an ecosystem, variation leads to biodiversity or a richness of species. Having many different species within an ecosystem, leads to stability within the food web.

10. 3 Scientists would first (step A) need to identify the gene that codes for the enzyme. They would then (step B) remove that gene from a healthy individual's DNA and (step C) insert that gene into bacterial DNA. The intent would be to have that bacterial DNA code for and direct synthesis of that enzyme (protein). Lastly (step D), the scientist would extract the enzyme from the bacteria for use as a treatment.

11. 2   Changes in gene frequency lead to variation within a population. Variation can lead to biological evolution if environmental conditions change. Genetic variation can be brought about by a change within the gene sequence, many times resulting in a mutation.

12. 4   The differences evidenced within the group of puppies from the same litter may be a result of chromosomes sorting during the process of meiosis and through recombination during sexual reproduction. New gene combinations can lead to variation within offspring. Such a variation might be coat or fur color.

13. 1   Respiration and photosynthesis are processes that maintain oxygen and carbon dioxide levels. Respiration, the energy producing process, uses oxygen and releases carbon dioxide to create the energy molecule, ATP. Photosynthesis uses carbon dioxide to produce sugars while releasing oxygen. Through each of these processes, carbon dioxide and oxygen levels in the atmosphere are maintained.

14. 3   In human development, the process starts with a single cell called a zygote (a fusion of the egg nucleus with sperm nucleus). The zygote goes through a series of mitotic divisions and differentiates into various types of tissue. Further development transforms these tissues into various organs which in turn work together within the developing fetus.

15. 3   This is a form of asexual reproduction known as vegetative propagation. In asexual reproduction, the offspring are genetically identical so the new plant will produce the same fruit (small and white) as the original plant.

16. 2   In order for a mutation to be inherited by an offspring, it must be carried by gametes or sex cells. A base substitution during meiosis, the gamete forming process, would create a mutation in the genetic information and be passed to offspring via that gamete during sexual reproduction.

17. 3   Proteins (*A*) are broken down into their building block known as amino acids (*B*). This process could occur as a result of digestion in organisms. During this process, molecules that are needed by the organism are made available from larger more complex molecules.

18. 4   Structures *A* (lungs) and *B* (kidneys) are both organs of excretion, responsible for the removal of metabolic wastes. Each organ requires an input of energy (ATP) to accomplish this process. The lungs excrete the waste product carbon dioxide and the kidneys excrete urea, salt, and water.

19. 2   Pathogenic organisms cause a disruption of homeostasis in humans, usually in the form of illness or disease. The effects of Salmonella bacteria on humans: fever, vomiting and diarrhea, characterize this bacteria as a pathogen.

20. 1   The AIDS virus HIV attacks or targets white blood cells which are directly responsible for fighting invading microbes or pathogens. The HIV virus weakens the human immune system and renders that person unable to fight against viruses or bacteria.

June 2009 Answer Key

21. 2   Guard cells surround and regulate the leaf openings called stomates. The atmospheric gases, oxygen and carbon dioxide enter and leave the plant through these stomata. By regulating the opening and closing of the stomata, the guard cells regulate oxygen and carbon dioxide levels.

22. 1   Bacteria reproduce asexually as shown in the diagram, with the offspring genetically identical to the parent cell. If an antibiotic is 95% effective in killing bacteria, the surviving 5% which are resistant would then be available to reproduce. Those bacteria have genes that make them resistant to antibiotics so when they asexually reproduce, the resistant gene is passed to the offspring. Over time bacteria will be more resistant to antibiotics.

23. 4   Minerals act as limiting factors for growth. In order to grow properly, plants must receive the correct types of minerals. If minerals were present in small amounts or lacking in an area, the area could not sustain a large population of plants. If the area were mineral rich, plant populations would be larger due to the availability of minerals necessary for growth.

24. 2   Competition occurs between organisms when they feed on the same nutrient or inhabit the same space. In this case, the chipmunk and squirrel are both competing for sunflower seeds as a food source. One species usually out competes the other based on an adaptation or characteristic that allows it to succeed.

25. 1   In all food chains and webs, as one moves up the chain(s), energy is released as heat. The heat energy released is a result of metabolic activity within that organism as it uses that food source.

26. 3   When members of a food chain or food web are removed from an ecosystem, there generally is a decrease in the stability of that ecosystem. If carnivores are removed, the species upon which they feed would increase, creating an imbalance in the levels of the food chain below them. For example, if wolves were removed, the populations of their food sources (deer and rabbits) would increase. This increase would result in overfeeding on plant sources and thereby create less food for other herbivores, causing a loss of stability in that ecosystem.

27. 4   The production and use of compost is an example of recycling. The plant materials used in compost contain nutrients such as organic substances and minerals. Decomposers break those materials down into a usable form such as fertilizer. The fertilizer then can be utilized by other plants or recycled into the ecosystem again.

28. 2   A chromosome contains long strands of DNA which is sequenced into many genes. These genes contain information that can be passed on during reproduction or used to synthesize proteins.

29. 4   Increasing levels of carbon dioxide can lead to global warming as a result of the greenhouse effect. An increase in atmospheric temperature as a result of increased carbon dioxide levels is of concern to environmentalists and scientists who continue to study its effects.

30. 2   Human activities over time have contributed to the depletion of finite resources. Knowing that finite resources cannot be replaced, humans must be aware of their actions and how they impact the environment and the resources found in that environment.

Part B-1

31. 2   The process of cloning results in an offspring (Dolly) that is genetically identical to the parent cell (body cell). Therefore the chromosome makeup of Dolly would be identical to Sheep A which donated the body cell.

32. 3   ATP is produced through the process of respiration within the mitochondria of the cell. Structure C within this cell represents the mitochondria. Remember that ATP is used as an energy molecule within living organisms.

33. 1   The maintenance of homeostasis allows the body to regulate or keep an internal balance. Secretions from glands allow the body to respond to any change that occurs and adjust to that change so that balance is restored. In this case, the pituitary, thyroid and adrenal glands and the pancreas all produce secretions to help maintain homeostasis.

34. 4   In this pond ecosystem, the catfish population act as consumers, feeding on rotifers and water fleas. By consuming rotifers and water fleas, the catfish control or keep those populations in check and maintain a dynamic equilibrium within the ecosystem. If the catfish did not control the rotifer and water flea populations, their numbers would become too large and diminish algae as a food source for other organisms.

35. 2   Structure B is a membrane protein. These proteins act as receptors of chemical signals based on the shapes of the protein and the chemical signal molecule. A specifically shaped signal molecule will initiate a cellular response when it binds to a particular protein.

36. 4   Organisms that have fast reproductive rates and live in changing environments would show fast evolutionary rates. Evolution reflects an organisms ability to adapt to changes in an environment. Certain genetic traits that allow organisms to survive and therefore adapt are passed reproductively from one generation to the next. If an organism has a rapid or fast reproductive rate, those adaptive genes will be passed faster thus allowing for a faster evolution rate.

37. 1   By using the Time Scale to the right of the diagram (Past → Present), many of the evolutionary pathways that represent descendants of organism *B* come to extinction (dead ends) well before the Present. The only two descendants that have successfully survived to the present are *Q* and *S* as those pathways reach the dotted line representing the Present time.

38. 1   In an energy pyramid, producers or plants (*A*) would have the greatest available energy. Remember that plants or producers are located at the bottom of the pyramid which contains the most energy. Producers gain their energy from the Sun by photosynthesis.

39. 4   Organism *C* (frog) and *E* (hawk) are carnivores. They feed on and obtain the energy from consumers below them in the energy pyramid. The hawk would feed on the snakes or frogs and the frogs would feed on the grasshoppers

40. 2   The evolution of differences in appearance between kit and red fox species occurred as each adapted to different environments. In each particular environment, size and fur color evolved through adaptation and survival allowing that trait to be passed on to offspring. For example, the lighter color fur of the desert living kit fox matches the lighter environment of the desert allowing the fox to camouflage itself as a predator as well as not absorbing as much light energy to prevent overheating in the desert.

41. 3   Algae and seaweed would be found near the ocean surface to maximize the amount of available sunlight absorbed for photosynthesis. Algae and seaweed are autotrophs or producers and use sunlight to produce energy through photosynthesis.

PART B-2

42. Answer: Both the dissolved oxygen and number of fish species decreased.

Explanation: Sewage water contains bacteria which would use oxygen in the water. As sewage (bacteria) increased, the amount of dissolved oxygen was greatly decreased. With less dissolved oxygen in the lake, fish populations were not able to sustain themselves. The number of populations decreased from 4 in 1950 (whitefish, trout, walleye and carp) to 2 in 1970 (walleye and carp)

43. The Temperature axis should be set up so that each line represents 10° C. The Amount of Gas axis should be set up so that each line represents 1 ml.

44.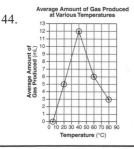

Average Amount of Gas Produced at Various Temperatures

45. 3 The maximum point both in the graph and the data chart for gas production is 40° C. In this experiment, the production of gas, a byproduct of cellular respiration, is used to measure the rate of respiration. The more gas produced, the more respiration occurred.

46. 2 Tube 3 contained the smallest amount of glucose. Since glucose serves as the energy source for respiration, the more respiration that takes place, the more glucose used. Tube 3 had the highest amount of gas indicating the highest amount respiration rate. Therefore it would have used the most glucose and would have smaller amounts than other tubes.

47.

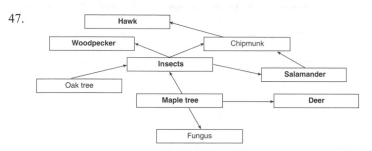

Explanation: Using the information in the observation chart, one can determine the food web in the forest ecosystem. Start with fungus at the bottom of the diagram, note that on 6/5 in the chart "fungus growing on maple tree" under Observed Feeding Relationships suggest that the term maple tree goes in the box directly over fungus on the diagram. On 6/11, "insect feeding on maple tree leaves" and on 6/5 "insect feeding on oak tree" both confirm that insect should go in the box directly over maple tree. Use chart to complete the diagram.

48. *Acceptable responses include but are not limited to:* maple Tree, tree, oak tree

Explanation: Trees are producers. They carry out photosynthesis using the Sun's energy. Organisms that feed on trees (consumers) gain that energy from the trees.

49. 1 Trees, being a living organism, are a biotic factor in the forest ecosystem. Photosynthesizing trees use water (an abiotic factor) to carry out that process. Trees absorb the raw material, water through their root systems from the ground within the forest environment.

50. *Acceptable responses include but are not limited to:*

2a Four legs ....   IV (Dog)
2b Eight legs......   II (Spider)

3a Fins present......   III (Fish)
3b Fins not present........ I (Earthworm)

Explanation: A dichotomous key is a tool used to identify organisms. At each step, two choices are offered that distinguish a difference between organisms. For example, From step 1a Legs present, we must look at the animals with legs (Dog and Spider). The task is then to find a difference such as different number of legs. At step 2 we identify these differences 2a - 4 legs and

2b - 8 legs and offer an identification of the animal. From step 1b, we follow the trait of legs not present and look at the remaining animals (Fish and Earthworm). The difference we might use could be the presence of fins. For 3a – Fins present –fish and for 3b – fins not present – earthworm.

51.

Name of Structure	Letter on Diagram	Function of Structure
ovary	C	produces gametes
uterus	D	site of internal development
placenta	B	transports oxygen directly to the embryo

Explanation: The function of producing gametes for the female reproductive system identifies the structure (1) as an ovary which is represented by (2) letter C. The uterus (D) functions to (3) provide a site for the developing embryo to implant. The structure labeled as B which serves to (4) transport oxygen to the embryo is the placenta.

PART C

52. Acceptable responses include but are not limited to:

Human Activities: Burning fossil fuels *or* dumping toxic wastes

Chemical released: CO_2, sulfur dioxide, nitrogen gas -or- heavy metals

Effect: Increase global warming *or* cause mutations

Human reduction: Use alternative fuels (solar, water or wind) Increase/enforce legislation to regulate toxic waste disposal

Explanation: Human actions can have long lasting effects on the environment. With increased use of fossil fuels as an energy source, humans are releasing increased amounts of CO_2 into the atmosphere. CO_2 increase leads to increasing temperatures on Earth known as global warming. One impact of global warming could be the extinction of organisms not able to adapt to climate change. Humans can reduce their fossil fuel use and therefore decrease CO_2 levels by developing and using alternative energy sources such as solar, wind and water energy.

53. Acceptable responses include but are not limited to:

Hypothesis:

Acidity: Acid rain will decrease the number of seeds that germinate.

Temperature: Exposing seeds to below average temperatures will slow down germination.

Amount of Water: Drought like conditions will decrease the rate of plant growth.

Control: Acidity- the control group is watered with water with a pH of 7 while experimental group is watered with water that is pH below 7.

Factors: Same soil, temperature, type of plant, water amounts, light or fertilizer

Independent variable: pH of water, temperature or amount of water.

Data Table

pH	Number of Seeds that Germinate

Explanation: The design of an experiment to test effects of environmental factors on plant growth could be focused on the effect of the acidity of precipitation. A hypothesis that could be tested would be that acid rain (pH below 7) will cause a decrease in the number of seeds that germinate (remember that a hypothesis should not be in the form of a question). For an experiment to be valid you must include a control as well as experimental groups. Experimental data will then be compared to control data to determine effect. In this experiment, the control would be water or rainfall of neutral pH (7). Experimental data would include varying pH's below 7 (in the acidic range). To insure that your measured effect is the result of one environmental factor, all other environmental factors should be kept the same for both control and experimental groups. Lastly, your data should be organized in a data chart. In this case there should be two columns: one that identifies the independent variable (pH) and one that identifies the effect on plant growth (Number of Seeds that Germinated). By analyzing the data, one can determine if the hypothesis should be accepted or rejected.

54. *Acceptable responses include but are not limited to:*

    No air pollution   *or*   burns less coal/oil   *or*   wind is a renewable resource

    Explanation: The use of windmills relies on the force of wind to drive turbines. This energy source uses no fossil fuels so there is no air pollution as a result of the burning of those fuels. Wind is considered to be a renewable resource as it cannot be used up

55. Acceptable responses include but are not limited to:

    Zebra mussels out compete native species for food, causing native species to decline in number.

    or

    Purple loosestrife has crowded out native species, leaving many native animals with much less available food, since they don't eat purple loosestrife.

    or

    Rabbits in Australia ate much of the vegetation that previously fed many Australian animals. Populations of many native species were drastically reduced, disrupting the ecosystem there.

    or

    Gypsy moths from Europe have overpopulated parts of the US, eating nearly all tree leaves, causing some trees to die, leaving little food for other native species.

    Explanation: Foreign species can alter the balance in ecosystems by becoming invasive within their new environment. Invasive species often have no natural predators and therefore overpopulate an area competing with native species for resources. The native species populations are reduced, creating a disruption in the food chain and thus ecosystem.

56. *Acceptable responses include but are not limited to:* Most flu viruses cause a runny nose and sore throat, while H5N1 can cause pneumonia.

or

The avian flu goes deeper into the lungs and can cause severe pneumonia.

or

The avian flu has a more severe effect on humans than most other flu viruses.

Explanation: In paragraph 2 of the passage, the effect of the H5N1 virus are described as different and much more serious than those of most virus strains.

57. Antibodies

Explanation: Antibodies are produced by white blood cells in response to foreign proteins or antigens that have entered the body. Antibodies function to mark viruses or infected cells for destruction or render viruses ineffective.

58. *Acceptable responses include but are not limited to:* Vaccines contain weakened or heat killed pathogens

Explanation: The purpose of a vaccine is to build immunity to a particular pathogen. By introducing weakened or dead pathogens into the body through a vaccine, a person will have an immune response that produces antibodies particular to that pathogen. When the person is exposed to that pathogen in the future, a rapid immune response can occur, effectively dealing with that pathogen, preventing sickness or disease.

59. Answer: Mutation

Explanation: Mutations are a sudden change or alteration of the genetic code. A change in the genetic code could result in a change in form or structure. By changing structure, a virus may be able to be spread from human to human by mimicking human protein shapes.

PART D

60.

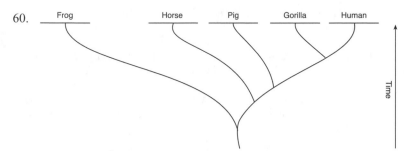

Explanation: One method of determining evolutionary position is to compare proteins or amino acid sequences of various organisms. Those organisms with a greater number of common amino acids are more closely related. Evolutionary relationships can be expressed in the form of an evolutionary tree which shows how organisms evolved and when new species appeared. For example, the gorilla has only one amino acid different from humans and would be placed on the branch closest the human. The frog which has the greatest number of differences would be placed on a branch farthest from the human to the far left.

61. Acceptable responses include but are not limited to:
Chemical similarities are more reliable than structural similarities.
or Electrophoresis shows chemical similarities which are more reliable.
or Many unrelated plants have a similar vein pattern.

Explanation: Gel electrophoresis is a process that separates DNA segments based on size. This separation creates a banding pattern in the gel that can be used to compare individual organisms to one another. Organisms with similar banding patterns are more closely related. This information can be used to develop evolutionary relationships between plants. Leaf vein patterns may not be unique to one species and therefore would not be as accurate as gel electrophoresis.

62. 2   This test for simple sugars such as glucose would be used to determine if the enzyme "digested" or broke down the soluble starch. Enzymes digest or break down starch into its component molecule, glucose. By using a glucose indicator, one could determine if starch had been broken down or digested with a positive test.

63. *Acceptable responses include but are not limited to:*
Pigment spot is below surface of the solvent
or Level of the solvent is too high

Explanation: Chromatography involves the use of a solvent to separate molecules. If the pigment spot on the chromatography paper is below the solvent surface, the pigment will dissolve into the solvent and not move up the paper with the solvent. By placing the pigment spot above the solvent, a separation of pigments will occur as the solvent moves up the paper, carrying molecules with it.

64. 1   By increasing the number of times the activity is repeated, the amount of data on which to base a conclusion is increased. This would lead to an increase in validity of the conclusion.

65. 73 beats/min

Explanation: To calculate the average of the resting pulse, add pulse data under Pulse Rate at Rest column for students A – E, then divide that total by the total number of students (5).

66. Acceptable responses include but are not limited to:
Heart – beats faster   *or*   Lungs –take in oxygen faster   *or*   Muscles – use energy faster or use more ATP.

Explanation: As activity increases, the rate of ATP use will increase. In order to supply more ATP to muscles, the rate of respiration must increase. Oxygen and glucose must be delivered to muscle cells to carry out this respiration. An increase in heart beat and oxygen taken into the lungs will provide these necessary molecules at a faster rate.

67. 2   The proper technique for adding salt to a wet mount slide involves using a dropper to introduce salt solution to one side of the cover slip. At the same time, place an absorbent cloth or paper on the other side of the cover slip to draw the water out. This will create a suction that will draw the salt water in on the other side into the area under the cover slip containing cells.

68. 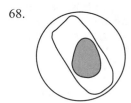   Explanation: The addition of salt water would create a situation where there was a lower concentration of water molecules outside of the cell than inside the cell. This would lead to osmosis of water out of the cell. (net movement from high to low concentration). The cell membrane would shrivel and pull away from the cell wall as a result. The plant cell wall will remain intact.

69. 4   To return cells to their original condition, water must move back into the cells. By exposing shriveled cells to distilled water, osmosis will occur, moving water into the cell because distilled water has a higher concentration of molecules than cell contents. The cells would be restored to their original condition with the cell membrane fitting closely to the cell wall.

70 3   By comparing the banding patterns of DNA of species X,Y and Z to species A, one can see that species Z has the most bands in common with A. These common bands suggest common DNA sequences and therefore a closer evolutionary relationship. Refer to the explanation in question #61 for a review of DNA electrophoresis.

71. 3   Natural selection is a process where organisms adapt to changes in their environment. Genetic variability due to mutations led to particular individuals being more successful on different islands. Those organisms survived while others did not. The Galapagos Islands provide varying habitats with varying food sources for many organisms leading to diversity through natural selection.

72. *Acceptable responses include but are not limited to:*
Large ground finches eat mainly plant food and large tree finches eat mainly animal food.
or   They do not compete for the same resources
or   They occupy different niches

Explanation: Using the diagram, find the Large ground finch and move directly in towards the center noting the type of beak and food type – mainly plant food. Do the same for the Large tree finch, noting the beak type and food type – mainly animal food. Since each bird feeds on different food types, they will not be competing for the same resources, allowing each bird to be successful on the island. Remember that two organisms cannot successfully occupy the same niche (role in the food web) at the same time. One will out compete the other.

June 2010
Part A

1. 4   Heterotrophs are organisms that acquire nutrients (preformed organic compounds) from their environment. Mushrooms rely on substances obtained from dead or decaying organisms to supply nutrients.

2. 2   Meat mainly consists of protein. After an organism eats the meat, its digestive system breaks that protein down into building blocks called amino acids. These amino acids are distributed to cells to be used to synthesize needed proteins.

3. 3   The circulatory system functions to distribute or transport needed nutrients to cells. Some of the substances transported by this system provide needed materials, like glucose and oxygen, necessary for the release of energy in the form of ATP.

4. 2   Hormones are designed to initiate a cellular response. In order to function, some hormones must bind to special membrane proteins called receptors. These receptors can usually be found on the cell membrane. The hormone's shape specifically fits with the receptor protein's shape to initiate the response.

5. 2   The excretion of most cellular wastes is taken care of by the cell membrane. This occurs either through the process of diffusion of waste molecules from high to low concentrations or by active transport across membranes.

6. 1   Vacuoles are designed to store materials that the cell may require at a later time. It can also store materials needed for defense, created as a by-product of metabolism, or waste product.

7. 2   Thymine, T, pairs with Adenine, A, so, there would also be 15% Adenine present in this DNA sample. The T – A amount would then total 30% leaving 70% to be made up of Cytosine, C, and Guanine, G. Dividing 70% by 2 to reflect the amounts of C and G individually, results in each being represented by 35%.

8. 1   Global warming is linked to a rise in temperatures. A direct result of these increased temperatures is the melting of the polar ice caps. When the icecaps melt, their size *decreases*.

9. 2 All cells, with the exception of sex cells, within a puppy contain the same genetic information. This genetic information was created when the zygote was formed. Through a process of cellular division, all cells that now make up the puppy will have identical DNA or genetic material. Remember that differentiation will "turn on" genes within cells and direct that cell to have a specific function/structure.

10. 2 The process of evolution is influenced by the passing of characteristics through gametes (sex cells) from one generation to another. If a change occurs that allows for survival due to a particular characteristic (adaptive trait) that change will be reflected within the genetic code of that organism. This change in the genetic code can only impact evolution if it is passed through gametes to future generations.

11. 1 Variation is one of the direct results of sexual reproduction, where two gametes recombine genetic information. Since genetic information is coming from two individuals, the resultant genetic information will be different from either parent. Agriculturists use the techniques of sexual reproduction and selective breeding to develop new varieties of vegetables.

12. 4 Genetic engineering includes a process where DNA is manipulated from one organism to another. In the diagram, a portion of DNA from a human has been inserted into bacterial DNA then placed in a bacterium. The bacterium will now carry out the instructions contained within that portion of human DNA.

13. 4 Many organisms use courtship behaviors to attract mates. Those organisms with the "flashiest" behavior are better able to attract mates, thereby insuring that their genetic information will be passed on. Since most behaviors have a genetic basis, the "flashy" behavior has evolved to promote reproductive success.

14. 1 According to the time line in the diagram, Species *B* and *C* are the only two species found in the present environment. Species *A*, *D* and *E* have all died out (become extinct). By definition, a species is reproductively isolated and will not breed with other species.

15. 1 In the diagram, the arrow represents fertilization, where the genetic information of the two gametes unites. After fertilization, the resulting zygote will have the complete set of genetic information for that species, getting half from the egg and half from the sperm.

16. 3   The expression of genetic material can be influenced by outside factors such as the environment. These factors can "turn on or turn off" genes so that their information will be expressed or not expressed. Twins, while having the same genetic makeup, could express different characteristics if they were exposed to different factors. For example, if one twin is exposed to sunlight and the other gets no exposure. The twin with sun exposure will exhibit a darkening of the skin that the other twin will not.

17. 4   After fertilization, the zygote with a complete set of chromosomes will begin to undergo rapid cell division (mitosis). These divisions will lead to differentiation, where the cells become specialized. This then leads to the development of an embryo.

18. 3   The production of gametes occurs in the ovary, represented by the structure numbered 3. Support of the fetus occurs in the uterus, represented by the structure numbered 5.

19. 3   The placenta provides a site where essential nutrients are transported between mother and fetus. The placenta also serves to transport waste products from the fetus to the mother who will excrete them. The placenta thereby serves as an exchange site between mother and fetus.

20. 1   In order for enzymes to function properly, they require a certain range of pH (specific to each enzyme). In the blood, if pH is not regulated, enzymes that clot blood may be impacted. Remember that an enzyme's shape determines its function. If pH changes or alters enzyme shape, it may not be able to function properly.

21. 1   Certain behaviors within a species are based on information found in genetic material. Even though young birds were isolated from other members of their species, they would still perform their behavior, such as nest building, because they are genetically programmed through inherited information from their parents. In this case, the environment would not play a part in their development of nest building skills.

22. 4   The human body's ability to regulate temperature is a homeostatic function. When this ability is altered by an injury, homeostasis is disrupted. In other words, the human body could begin to experience illness or even death due to high temperature brought on by the lack of sweating ability.

23. 4   Decomposers break down materials into useful nutrients. These
        nutrients are then available to plants for use in carrying out
        photosynthesis as well as respiration. Inorganic nutrients like
        minerals can be recycled back into plants by the action of these
        decomposers. Bacteria and some fungi act as important
        decomposers in ecosystems.

24. 2   Manatees acting as herbivores in an aquatic ecosystem feed on water
        plants or producers. If the manatees become extinct, then there
        would be less herbivores feeding on water plants. These producers
        would then increase in population or become more abundant.

25. 1   The loss of habitat through destruction by humans or natural means
        would directly decrease biodiversity in that area. Biodiversity is the
        number or richness of species in a given area. With loss of habitat,
        organisms would have no place to live. Either they would die out or
        move, decreasing the number of organisms in that ecosystem.

26. 3   By cutting down all of the trees for lumber, the forest would lose
        most of its producers. Producers are the base for a food web or
        energy pyramid in an ecosystem. If trees (producers) are removed,
        the stability of that ecosystem is decreased because these organisms
        provide energy for the rest of the organism in that ecosystem.

27. 2   Renewable resources can be replenished and then reused. By using
        renewable resources, humans can positively impact their local
        environments by not using up resources that cannot be replaced.

28. 3   Evolution is a change in a population over time as a result of a
        change in the environment. Climate change in a region may allow
        for change in plant populations. Those with favorable characteristics
        would survive, perhaps leading to new varieties of plants. This region
        undergoing climate changes would show more evidence of evolution
        based on more plant varieties.

29. 4   The hardwood forest or climax stage would be the most stable. Based
        on the numbers and types of producers, consumers and decomposers
        that are best fit for the environment at this stage, it will remain stable
        until there is either a natural or man made disturbance.

30. 1   A trade off looks at short term vs. long-term factors. While the solar
        panels would be expensive to purchase in the short term, the long-
        term benefits would be reduced fuel costs (less oil) and lower taxes
        due to a tax rebate. Overall, the homeowner saves money as time
        passes.

PART B -1

31. 3   Using the information provided in the diagram, there are 5 cells within the 1 mm field of view. 1 mm = 1000 μm, so divide 1000 μm by 5 cells to get the answer of 200 μm.

32. 3   In the diagram, species B's numbers are increasing while species A's numbers are decreasing. Based on these numbers, one could offer the explanation that species B is better adapted to its environment based on characteristics or traits that allow it to survive and reproduce successfully, thus, increasing its numbers.

33. 4   In a valid experiment, you should have a control group which is not receiving the supplement so as to determine if it is the supplement that is effective against the flu. You would test the two groups (A and B) with the supplement, and add a third group that gets nothing or a placebo to act as the control.

34. 1   Row 1 identifies each cause of mutation. Mutation A is a deletion, note that the second A in the normal sequence has been deleted (or left out) of mutation. In mutation B, a substitution has taken place, C for G. For mutation C, a G is inserted between the two T's in the normal sequence.

35. 1   Organisms in level A represent the highest level of consumers. In the energy pyramid, organisms receive their energy from the level directly below them. Thus, level A receives its energy from level B.

36. 4   Level D represents producers. Producers are autotrophs that receive their energy from the Sun. The Sun is not part of the ecosystem – it is an outside energy source.

37. 4   A gene is located on a chromosome which can be found in a nucleus. Based on this, the gene would be the smallest in size and the nucleus would be the largest in size.

38. 2   In the diagram there are more X's outside of the cell than inside. Moving molecules out of the cell would require going against the gradient from low to high concentration. In order to move the X molecules out, the energy molecule, ATP, would be required. This is known as active transport.

39. 4 In the aquarium setting, there are producers, consumers and microorganisms that will act as decomposers. If this is maintained in natural light, all organisms or parts of this ecosystem would cycle energy and materials with no outside help. Light energy → producers → food energy → consumers → decomposers → recycle nutrient to producers.

40. 2 Secondhand smoke during pregnancy slows fetal growth based on data from the table. In nonsmoking couples, the fetus averaged 3.2 kg but in couples where the husband smoked, the fetus averaged 2.9 kg, a decrease in average birth weight. This suggests that secondhand smoke affects fetal growth negatively.

41. 4 Soil nutrients are used by plants during growth and need to be replenished into the soil for each growing season. Without replenishing them, corn growth would be limited in the future. All other given choices are renewable.

42. 2 When building the Aswan Dam for positive impact to humans (irrigation and energy), the planners did not foresee that the change in flow of the Nile would negatively influence human health. By changing the habitat, the pupae of parasitic larvae increased, causing an increase in infection within the human population.

43. 1 The snail serves as a host to the larvae of the fluke. A host serves as either a food source or as a habitat for parasites. Parasites usually harm their hosts causing disease, illness, or even death.

PART B-2

44. Carbon dioxide, CO_2

Acceptable responses include but are not limited to:
– driving automobiles
– burning fossil fuels
– deforestation
– respiration

Explanation: Based on % abundance, carbon dioxide is the most abundant greenhouse gas, making up 99.438% of these gases. Greenhouse gases result in a warming of the Earth's atmosphere. Human activities such as driving cars, burning fossil fuels, deforestation (burning forests) and respiration all release carbon dioxide into the atmosphere, increasing the possibility of global warming.

45. Acceptable responses include but are not limited to:
   - Organisms may contain parasites or diseases which could spread to other organisms in this country.
   - Imported species may out compete native or domestic species for food or habitat.
   - there are no natural predators

   Explanation: If a foreign species or organism is introduced into a new country, it may be harmful to organisms native to that country. For example, a plant may bring a disease that could potentially wipe out United States' crops because they have no natural defenses. Likewise, a plant could out compete native plants for nutrients and space causing those species to decrease. Being foreign also might mean that there are no natural predators to keep that organism's population in check, so it may increase in numbers rapidly, also competing for resources.

46 –47

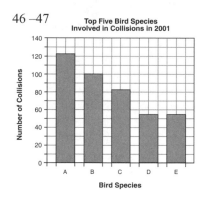

Explanation: For an appropriate scale, the range of collision values go from 55 to 123, so assign a value of 10 to each line on the Number of Collisions axis. You may label every other line, as done in the graph. Locate the appropriate value for each bird type (*A*, *B*, *C* or *D*) and mark a bar at the top of the column for that bird. Fill in the area under that column. For example, for bird *A*, the value is 123. Make a bar across the space over *A* at collision value 123 and shade in that column under the bar.

48. Answer: No

   Acceptable responses include but are not limited to:
   - only 50% occur in airfields

   Explanation: Collisions are not limited to airfields. Based on the article, 50% occur in other areas. Birds fly in all areas, not just airfields, so there would always be the possibility of having a bird-plane collision outside those airfields.

49. Acceptable responses include but are not limited to:
   - migration occurs during these months
   - birds feed in these areas at this time

   Explanation: August, September and October are months when birds are migrating to warmer climates for the winter season. Because birds tend to migrate in large numbers, there is a greater possibility of a collision with a plane. Birds could also be actively feeding in preparation for winter in these areas, leading to more collisions.

June 2010 Answer Key

50. Acceptable responses include but are not limited to:
    – Transpiration helps cool plants off on hot days.
    – During transpiration, stomata are open allowing for gas exchange.

    Explanation: Transpiration is the loss of water through evaporation from stomata located on the leaves of a plant. Evaporation is a cooling process, so through evaporation, stomata on leaves can regulate a plant's temperature. This can prevent harmful high temperatures, which can negatively affect proteins within the plant. Another benefit from transpiration is the exchange of gases such as carbon dioxide and oxygen through the stomata. Carbon dioxide is necessary for the energy producing process, photosynthesis. By having the stomata open, carbon dioxide is available to be converted by light energy into organic compounds like sugars.

51. Acceptable responses include but are not limited to:
    oxygen *or* carbon dioxide *or* water vapor

    Explanation: Exchange of gases through stomata openings allows for oxygen to exit the plant and enter the atmosphere. Oxygen is necessary for cellular respiration in all living organisms. Carbon dioxide can enter the leaf through stomata and be made available for photosynthesis. Carbon dioxide is converted into sugars during the process of photosynthesis. Water vapor is also exchanged in and out of the plant through stomata.

52. Answer: Guard cells

    Explanation: Guard cells surround stomata openings. When water and nutrients enter guard cells, they swell, causing the stomata to open. Likewise, if water/nutrients move out of guard cells, the opening closes. In this way, guard cells regulate the opening and closing of stomata.

53. Acceptable responses include but are not limited to:
    – if enzyme shape changes, its function change also
    – key activity may slow down or stop
    – enzyme shape determines enzyme function

    Explanation: Enzymes are proteins that allow reactions to proceed efficiently. They have specific shapes that determine their function. If an enzyme's shape is changed, it cannot carry out its function efficiently and therefore will affect the activity or metabolism of the plant. The plant may not grow or it might become diseased and die.

54. Acceptable responses include but are not limited to:
    disease *or* harsh winter *or* decreased food supply *or* predators

    Explanation: At the beginning of time *X*, the population number was high. This large population number could lead to crowding where food supplies may be limited and therefore increase the possibility of disease

(continued on the next page)

or stress. There could also be an increase in predation on the coyotes. For example, a state or county could establish a bounty for dead coyotes, thus reducing the population.

55. Answer: *Type of Gamete* — Female *or* ZW *or* egg

Supporting answer: Acceptable responses include but are not limited to:
– Female gametes have one of two different types of sex chromosomes.
– The egg may contain one of two different types of sex chromosomes.

Explanation: With birds, the female gamete determines the sex of the offspring. If the female gamete contains a Z, the offspring will be a male, and if the gamete contains a W, the offspring will be a female. In humans, the male gamete determines the sex of the offspring by providing either an X or a Y chromosome. If the gamete contains an X, the offspring will be a female, and if the gamete contains a Y, the offspring is male.

Part C

56. Acceptable responses include but are not limited to:
Hypothesis:
  – Rats given the drug will have an increase in BDNF in their blood.
  – The drug affects memory function in rats.
Control group:
  – Control group will not get the drug, experimental group does.
  – Control group will get a placebo.
Factors:
  – Each group should have same number of rats.
  – rats are the same age
  – rats are kept in same conditions
  – rats are given the same food
Dependent variable:
  – amount of BDNF found in nervous tissue
  – ability to form/store memories

Sample answer: To test the *hypothesis* that rats, when given the new drug, will show an increase of BDNF in their blood, an experiment will be conducted with two rat groups. One group, the *control group*, will not be given the drug; the second group will receive the new drug and be known as the experimental group. All other *factors* must remain constant within the experiment, such as age of rats, type and number of rats, as well as their food and conditions in which they live. This is to make sure that the only variable we are testing is taking the drug or not taking the drug. To measure if the hypothesis is valid, the experimental data or measureable *dependent variable* should reflect that the rats have an increase in BDNF in their blood or that they show an increase in their ability to store or form memories.

57. Acceptable responses include but are not limited to:
    – The above average leg strength trait would increase in frequency because the rabbits with the stronger legs would be more likely to get away from predators.

    Explanation: If rabbits with extra strength legs can escape predators, they will be available to reproduce and pass that favorable leg trait on to their offspring. Rabbits without the extra strength trait may be preyed upon more often and not be as reproductively successful. Over time, the successful extra strength leg trait frequency will increase within a population.

58. Acceptable responses include but are not limited to:
    – these rabbits will start to decrease in numbers
    – they will be eaten by predators

    Explanation: Referring to the answer in question 57, the rabbits without the leg strength trait will be preyed upon more easily and will decrease in number.

59. Acceptable responses include but are not limited to:
    – The frequency of the trait might decrease because poor eyesight might be more of a disadvantage than the leg strength is an advantage.
    – Frequency will likely decrease because the rabbits won't see well enough to escape from the predator.
    – Frequency of the trait will remain the same because leg strength advantage is cancelled out by poor eyesight disadvantage.

    Explanation: Even though the rabbits with extra leg strength have an advantage escaping from predation, they may not see predators soon enough to get away. If these rabbits are caught and killed, they will be unable to reproduce and pass on the leg trait. The frequency of that trait will decrease within the population.

60. Acceptable responses include but are not limited to:
    Gamete event: mutation   *or*   change in DNA code
    Overuse: – Antibiotics kill only nonresistant strains. Resistant bacteria survive and reproduce.
    – Overuse of antibiotics leads to selection for resistant strains of bacteria.

    Explanation: Within a population of bacteria, there will be variation, which could be the result of a mutation within the DNA code. Certain bacteria may have genes which allow them to be resistant to antibiotic action, while others lack the gene. If the bacteria are subjected to antibiotics, those bacteria that lack the gene do not survive, while those with the resistant gene do survive. The surviving resistant bacteria can then reproduce and pass on that resistant gene. With each use of antibiotics, there will be more surviving resistant bacteria. The bacteria population will eventually become antibiotic resistant.

61. Acceptable responses include but are not limited to:
better health awareness *or* better medical care *or* better sanitation *or* new medical procedures

Explanation: Life expectancy has increased in the United States due to the discovery and use of new medicines that prevent or decrease diseases within the population. New medical procedures also help to prolong life expectancy. Humans are also more educated about their health and many maintain healthy life styles. Better sanitation practices also prevent the spread of disease causing pathogens, increasing human life expectancy.

62. Acceptable responses include but are not limited to:
– Humans need for space, food or resources often harm or interfere with other species.
– As humans build more houses, roads, malls, etc., animal habitats are destroyed.
– Increased populations increase competition with other species for limited resources.

Explanation: An increase in human population can have a negative affect on other species. More humans could mean an increase in competition for available resources and the need for more housing. With a need for more space, humans would move into animal habitats or destroy habitats for lumber to build those homes.

63. Answer: Organelle *X* is a mitochondria.

Acceptable responses include but are not limited to:
Process: – respiration /cellular respiration
        – aerobic respiration
        – releases energy (ATP)
Raw materials:
        – Sugar/glucose
        – oxygen
Molecule Produced and Importance:
        – ATP – provides energy for life processes
        – water – important for chemical reaction
        – carbon dioxide – waste product that must be removed to maintain homeostasis

Explanation: Organelle *X* is a mitochondria, which are cellular organelles that produce ATP, an energy molecule as a result of cellular respiration. Glucose and oxygen are raw materials necessary for the process of respiration to take place. Carbon dioxide is produced as a waste product and must be removed from the cell to maintain homeostasis both within the cell and within the organism.

June 2010
Answer Key

64. Acceptable responses include but are not limited to:
– The wolf population would decrease.

Supporting explanation: If the amount of vegetation decreases, the number of wolves would also decrease. With less vegetation, herbivores would decrease in number, resulting in less available food for the wolves. As a result, the wolf population would decrease due to this decreased food supply.

Part D

65. Answer: Species A and *B* are most closely related.

Supporting Explanation:
Acceptable responses include but are not limited to:
– They are most closely related because there is only one difference
  in the sequence.
– they have the most in common

Explanation: Organisms that are closely related have amino acid sequences that are similar or very close in sequence. Species A and B have only one amino acid difference in the sequence shown. At the fourth location, species A has Ser, while species B has Cys. Therefore, they are the most closely related. Species C has two differences in sequence compared to A and three compared to B.

66. Acceptable responses include but are not limited to:
– as she continued to use her hand, her muscles became fatigued
– wastes build up
– ATP was used up

Explanation: Muscle activity requires the energy molecule, ATP. If muscle activity continues beyond the available supply of ATP, there will be a reduction in the amount of activity that the muscle can do. As a result of muscle activity, waste products build up and can cause muscle fatigue, a situation where the muscles cannot maintain their full level of activity. Thus, less muscle activity will occur.

67. Answer: Diffusion *or* Passive Transport

Explanation: Diffusion or passive transport is the movement of molecules from an area of high concentration to an area of low concentration. In the diagram, the solution inside the dialysis tubing contains 10% glucose, making it more concentrated than the water containing 0% glucose. So the net movement of glucose will be from the inside of the artificial cell (10%) to the outside beaker (0%).

68. Acceptable responses include but are not limited to:
– According to diagram 3, C should look different from A and B.
– Stem cross-sections in diagram 1 show that A, B and C have similar stem structures, indicating that they are most likely related.
– diagram 3 shows only A and B as being closely related

Explanation: Diagram 3 shows the evolutionary relationship based on DNA data from diagram 2. The DNA data shows that species A and B share more gel electrophoresis bands than they do with species C. This suggests that species A and B are more closely related and would be closer on an evolutionary tree than they would be with species C. The information in diagram 1 doesn't support this, as all of the cross sections are similar.

69. Acceptable responses include but are not limited to:
– Species A and B have the most bands in common.

Explanation: Species that are closely related will share similar DNA sequences. When the DNA is subjected to gel electrophoresis, species with close evolutionary links will have more bands in common. In diagram 2, species A and B share 3 bands in common, while C shares one band in common with B and 2 with A.

70. 2   Gel electrophoresis is a process that separates DNA fragments that have been cut up by restriction enzymes. The DNA is placed in a gel and an electric current, is applied. The DNA migrates through the gel as a result of the electric current with the smallest fragments moving the farthest. Banding patterns occur and can be analyzed to look for similarities between samples (bands in common).

71. Acceptable responses include but are not limited to:
– DNA analysis is more reliable. The more similar the DNA, the closer the relationship
– Organisms can have similar features, but DNA coding for these features can be very different.
– DNA analysis might reveal the actual genetic makeup.

Explanation: DNA analysis can reflect the actual genetic makeup of organisms. Gel electrophoresis can show shared DNA sequences between organisms. While microscopic views of stem cross sections may show similarities, they are not as reliable as comparing DNA to determine evolutionary relationships.

72.

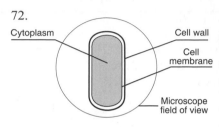

Cytoplasm

Cell wall

Cell membrane

Microscope field of view

Explanation: In plant cells (red onion), the cell wall is the outer wall surrounding the cell. The cell membrane, which regulates movement in and out of the cell, is located directly inside of the cell wall. The cytoplasm is the liquid matrix that fills the inside of the cell and is represented by the shaded gray area.

73. Acceptable responses include but are not limited to:
   – Place saltwater solution on one side of the coverslip. Then draw that saltwater solution under the coverslip by placing a piece of paper towel on the opposite side.
   – Place a drop of saltwater solution on one side of coverslip and a paper towel on the other.

Explanation: To add saltwater to a slide without removing the coverslip, place a drop of saltwater on one side of the coverslip. Put a paper towel on the opposite side of the coverslip. The paper towel will draw or soak up water from under the coverslip, creating a suction that will pull the saltwater from the opposite side under the coverslip.

74.

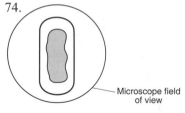

Microscope field of view

Explanation: When saltwater is added, the cell loses water and the cell membrane will pull in from the cell wall. This occurs because saltwater contains a lower concentration of water than the cytoplasm. Water will diffuse out of the cell from a higher to a lower concentration of water.

75. 1   Specific enzymes called restriction enzymes are used by scientists to cut DNA samples into smaller segments or fragments. These enzymes cut the DNA at specific sites, creating specific sized fragments of known length. These DNA fragments can then be used in gel electrophoresis for DNA analysis.

76. 2   The different beaks of finches are a result of natural selection. Organisms with favorable characteristics survive and reproduce, passing on these favorable traits. On each Galapagos Island, the specific environment and food type selects for specific favorable beak type, thus different beaks are observed on different islands.

77. 2   A niche is the role that an organism fills in an ecosystem. Living in different areas, the two birds do not have to compete for prey. When their territories overlap, there will be competition. By changing their niche to night feeders, one species adapts to avoid competition and insure survival.

LIVING ENVIRONMENT

CONCEPTS, RELATIONSHIPS, AND QUESTIONS BY TOPIC REFERENCE

1. Cells are the basic unit of structure and can be differentiated to perform many different functions.

2. Exceptions to the Cell Theory include: the first cell, viruses and chloroplasts and mitochondria.

3. Organisms are classified based upon structural and evolutionary relationships.

4. All living organisms must carry out all life processes or functions in order to survive.

5. Homeostasis is the maintenance of internal balance within an organism.

6. Organic compounds are composed of building blocks which are essential for living organisms. They include:
    Carbohydrates – composed of simple sugars
    Proteins   – composed of amino acids
    Lipids   – composed of fatty acids and glycerol

7. Organic compounds can be identified by the use of indicators such as:
    Lugol's solution   – identifies starch
    Benedict's solution – identifies simple sugars
    Biuret's solution   – identifies proteins

8. Enzymes are organic catalysts that speed up a chemical reaction but are unchanged by that reaction.

9.

    Enzyme   Substrate   Enzyme Substrate Complex   Enzyme   Product

10. Rates of enzyme activity can be influenced by temperature, pH and the amounts of enzyme or substrate.

11. Autotrophs produce organic molecules from inorganic materials, generally, through photosynthesis.

12. Heterotrophs must obtain pre-made organic molecules for nutrition.

13. Red, blue and violet wavelengths of visible light are optimum for photosynthesis.

14. Photosynthesis involves a process where light energy is changed to chemical energy which is stored in food molecules.

Concepts and Relationships

15. The cell membrane (plasma membrane) regulates homeostasis within a cell by selectively allowing materials in or out.

16. Active transport involves the use of energy to move materials from areas of low concentration to areas of higher concentration.

17. Diffusion and osmosis require no energy for movement, relying on a concentration gradient (high to low concentration).

18. Cellular respiration results in the conversion of chemical energy in food (glucose) into a stored form of energy known as ATP.

19. Aerobic respiration takes place within the mitochondria, requires oxygen and results in 36 ATPs being produced

20. The human kidney regulates internal water balance

21. Metabolic waste products include: CO_2, water, salts and nitrogenous wastes such as ammonia, urea and uric acid

22. White blood cells are involved in body defense by engulfing bacteria and and other foreign material through phagocytosis.

23. Immunity to disease relies on having specific antibodies that attack a particular antigen (foreign material).

24. Active immunity occurs through direct contact with the disease–causing organism or antigen–or–by having a vaccination which contains killed or weakened disease.

25. Passive immunity is temporary and is achieved by introducing antibodies into the body.

26. HIV virus attacks and weakens the immune system by disabling white blood cells.

27. Illness and disease cause a disruption in homeostasis.

28. Variability within a species is accomplished through sexual reproduction, crossing over and mutation.

29. Individual characteristics are inherited as a result of the transmission of genes through meiosis and sexual reproduction.

30. Changes in genes occur as a result of mutation, a change in the chromosomal number or an alteration of the genetic code.

Concepts and Relationships
          Living Environment

31. Mutagenic agents include: radiation (X-ray and UV) and chemicals (pesticides and pollutants).

32. Environmental factors, such as temperature, can influence gene expression or development. Genes can be turned on or off.

33. Artificial selection is the practice of choosing desirable traits and maintaining them within a species through selective breeding.

34. Detection of genetic disorders can be performed by karyotyping, DNA profiling and amniocentesis.

35. Genetic engineering can be accomplished through selective breeding, insertion of human DNA into bacterial DNA (gene splicing), cloning and gene therapy.

36. Evidence for evolutionary theory includes: the geologic record, biochemical similarities(DNA), comparative anatomy and comparative embryology.

37. Natural selection as suggested by Darwin occurs when traits which promote survival are passed on to offspring.

38. Resistance to chemicals such as insecticides and antibiotics occurs as a result of the evolutionary process of natural selection

39. Abiotic (non-living) ecological factors include: light, temperature, moisture, inorganic materials, gases (O_2 and CO_2) and soil.

40. Symbiotic relationships include: commensalism (+,0), mutualism (+,+), and parasitism (+,–).

41. Energy within an ecosystem flows up through the food chain from autotrophs to heterotrophs.

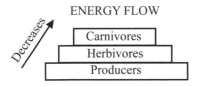
ENERGY FLOW

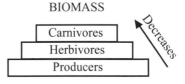
BIOMASS

Concepts and
Relationships

Carbon Cycle

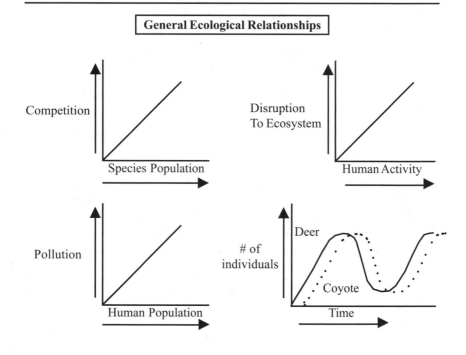

```
                    ┌─────────────────┐
                    │  Carbon Doxide  │
                    │     Water       │
                    └─────────────────┘
```

Carbon Doxide Water → Organic compounds made by autotrophs through photosynthesis

Resiration releases

Organic compounds made by autotrophs through photosynthesis → Organic compounds taken in by heterotrophs

Oxygen

General Ecological Relationships

Competition — Species Population

Disruption To Ecosystem — Human Activity

Pollution — Human Population

# of individuals — Time

Deer

Coyote

REGENTS QUESTIONS BY TOPIC

Scientific Methods and Laboratory Skills
June 2007 – 31, 32, 35, 49, 50, 51, 52, 56, 62, 63
June 2008 – 31, 33, 37, 43, 44, 51, 52, 63, 69, 77, 79
June 2009 – 43, 44, 45, 50, 53, 62, 63, 64, 65, 67, 68, 69
June 2010 – 31, 33, 46, 47, 56

Cells and Biological Processes
June 2007 – 2, 9, 19, 26, 43, 45, 71, 72, 73
June 2008 – 1, 4, 5, 9, 10, 35, 36, 49, 64, 76, 78, 80
June 2009 – 3, 4, 17, 22, 32, 35
June 2010 – 2, 5, 6, 20, 38, 53, 63, 67, 72, 73, 74

Energy – Photosynthesis and Respiration
June 2007 – 20, 24, 25, 53
June 2008 – 15, 19, 34
June 2009 – 13, 21, 46
June 2010 – 66

Human Physiology
June 2007 – 3, 4, 5, 15, 16, 17, 18, 21, 39, 40, 41, 55, 64
June 2008 – 17, 18, 20, 22, 23, 53, 54, 55, 56, 57, 70
June 2009 – 2, 6, 14, 18, 19, 20, 33, 51, 56, 57, 58, 59, 66
June 2010 – 3, 4, 17, 18, 19, 22, 40, 55, 61

Genetics
June 2007 – 6, 7, 8, 10, 11, 23, 33, 59, 65, 66, 67
June 2008 – 3, 6, 7, 11, 12, 16, 21, 32, 38, 50, 68
June 2009 – 5, 8, 10, 11, 12, 15, 16, 28, 31
June 2010 – 7, 9, 11, 12, 13, 15, 16, 21, 34, 37, 57, 58, 59, 60, 65, 70, 71, 7

Evolution
June 2007 – 12, 13, 14, 22, 37, 68, 69, 70
June 2008 – 8, 13, 14, 40, 41, 71, 72, 73
June 2009 – 36, 37, 40, 61, 70, 71, 72
June 2010 – 10, 14, 28, 68, 69, 76

Ecology
June 2007 – 27, 28, 29, 30, 34, 36, 38, 42, 46, 47, 48, 57, 58, 60, 61
June 2008 – 2, 24, 25, 26, 27, 28, 29, 30, 39, 42, 45, 46, 47, 48, 49, 58, 59, 60, 61, 62, 65, 66, 67, 74, 75
June 2009 – 1, 7, 9, 23, 24, 25, 26, 27, 29, 30, 34, 38, 39, 41, 42, 48, 49, 52, 54, 55
June 2010 – 1, 8, 23, 24, 25, 26, 27, 29, 30, 32, 35, 36, 39, 41, 42, 43, 44, 45, 48, 49, 50, 51, 52, 54, 62, 64, 77

Correlations